Treatments

Treatments

Language, Politics, and the Culture of Illness

Lisa Diedrich

University of Minnesota Press
Minneapolis
London

An early version of chapter 1 was published as "Space, Language, and Death in Twentieth-Century Memoirs of Tuberculosis," *Studies in the Social Sciences* 38 (July 2005): 45–62. Portions of chapter 4 were published in *Journal of Medical Humanities* (Winter 2005). Portions of chapters 4 and 5 were published in "Breaking Down: A Phenomenology of Disability," *Literature and Medicine* 20, no. 2 (Fall 2001) and in "Without Us All Told: Paul Monette's Vigilant Witnessing of the AIDS Crisis," *Literature and Medicine* 23, no. 1 (Spring 2004); reprinted with permission from The Johns Hopkins University Press. Portions of the conclusion were published in "A Bioethics of Failure: Anti-heroic Cancer Narratives," in Margrit Shildrick and Roxanne Mykitiuk, eds., *Ethics of the Body: Postconventional Challenges* (Cambridge, Mass.: The MIT Press, 2005), 135–52; reprinted with permission from The MIT Press.

Poetry excerpts from *Love Alone*, by Paul Monette, copyright 1988 by the author and reprinted by permission of St. Martin's Press, L.L.C.

Published by the University of Minnesota Press
111 Third Avenue South, Suite 290
Minneapolis, MN 55401-2520
http://www.upress.umn.edu

Library of Congress Cataloging-in-Publication Data

Diedrich, Lisa.
Treatments : language, politics, and the culture of illness / Lisa Diedrich.
p. ; cm.
Includes bibliographical references and index.
ISBN 10: 0-8166-4697-X (hc : alk. paper) — ISBN 10: 0-8166-4698-8 (pb : alk. paper)
ISBN 13: 978-0-8166-4697-5 (hc : alk. paper) — ISBN 13: 978-0-8166-4698-2 (pb : alk. paper)
1. Sick—Psychology. 2. Patients' writings—History and criticism. 3. Diseases and literature. I. Title.
[DNLM: 1. Medicine in Literature. 2. Attitude to Death. 3. Attitude to Health. 4. Autobiography. WZ 330 D5585t 2007]
R726.5.D54 2007
610—dc22 2007001734

Printed in the United States of America on acid-free paper

The University of Minnesota is an equal-opportunity educator and employer.

Contents

Introduction
Doing Treatments

This is a book about bodies, language, and death; it is about illness, culture, and politics, and the discourses and practices of medicine, literature, and philosophy. I have chosen to look at these broad themes as they come together in a particular object, contemporary memoirs about illness. In *Treatments*, I investigate the sociohistorical contexts in which these illness narratives[1] are being written and read, as well as the subjective experience of the ill and dying body as described in these narratives. I consider what kinds of cultural work illness narratives do in contemporary Western culture: for those who write them, for those who read them, and for those institutions (literary, medical, cultural) that they become a part of or remain outside of.[2] I explore the ways that illness narratives can be read as symptomatic texts of our time in at least two respects: as texts that literally describe symptoms (and struggle with finding a form to describe the affective and physical experience of symptoms), and as texts that describe illness as an event that goes beyond any particular individual's experience and account of it, reflecting wider cultural categories, including race, gender, class, and sexuality.

My method for reading illness narratives is interdisciplinary. In her book *Ghostly Matters*, Avery Gordon quotes Roland Barthes on interdisciplinarity: "Interdisciplinary work, so much discussed these days, is not about confronting already constituted disciplines (none of which, in fact, is willing to let itself go). To do something interdisciplinary it's not enough to choose a 'subject' (a theme) and gather around it two or three sciences. Interdisciplinarity consists in creating a new object that belongs to no one" (Gordon 1997, 7). The interdisciplinary tool that I use is the theory/method that I call *treatments*, a term that has, according to the *OED*, multiple meanings, including: "the process or manner of behaving towards or dealing with a person or thing"; "the application of medical care or attention to a patient, ailment, etc."; "a manner or instance of dealing with a subject or work of literature, art, etc."; and, perhaps most important, "discussion or arrangement of terms, negotiation."[3] Taking into account all of these definitions, in *Treatments* I negotiate the following questions: What sort of *subject* is formed in the *practice* of writing memoir in general, and illness narratives in particular? What sorts of *knowledges* are articulated in such writing? How are these knowledges different from expert medical knowledges, and are they capable of transforming expert medical knowledges? How does *language* both capture and fail to capture the "scenes of loss"[4] portrayed in illness narratives? And, finally, what sort of *ethics* emerges out of such scenes of loss and the attempts to capture them in writing?

The new object that is created through the various case studies treated here is the experience and event of illness, which, within the context of illness narratives, crosses multiple domains—literary, philosophical, cultural, political, medical—and does not belong to the "experts" of these domains. Because it crosses these multiple domains, *Treatments* is a project that relates to other projects in three emergent interdisciplinary fields: medical humanities, science studies, and disability studies. I hope *Treatments* will also appeal to scholars across the disciplines of literature, the social sciences, and medicine because I am very much interested in negotiating conversations and projects across these disciplinary divides. I think there is also an appeal to a more general audience, because, although I do work with theory quite extensively, I believe that theory emerges out of experiences and events in the real world, abstracts them, and then becomes a prism through which we can view these experiences and events anew. The abstraction that theory performs does not move us beyond the world of "real" experiences and events, but rather necessitates that we view them otherwise.

Throughout *Treatments*, I am concerned with the relationship of the memoir form to transformation (of selves, language, and institutions) and conformation (to established narrative and institutional structures and practices). What is it about the writing and reading of memoir that is potentially transformative, and how, at the same time, does memoir conform to and participate in hegemonic political and aesthetic structures and idioms? By engaging with these contemporary cultural debates, I hope to situate the more specific object of my inquiry—autobiographical accounts about the experience of illness—into a wider sociocultural context. In my analysis I use three theoretical frames—poststructuralism, phenomenology, and psychoanalysis—that appear, at least on the surface, to be incompatible, although they might also be said to come together or be linked under the more expansive and interdisciplinary rubric of feminist theory. My intention is not to show that any one of these frames is the best theoretical lens through which to read illness narratives, but rather to use them with and against each other in order to delineate the ways in which illness narratives are both effective and affective histories; that is, histories that are attentive both to the rhetorics and practices of politics as well as to the poetics and practices of suffering.

"Victim Art" and "Experience"

In "Discussing the Undiscussable," a brief article that appeared in the *New Yorker* in the mid 1990s, dance critic Arlene Croce trenchantly argues against what she calls "victim art." In her polemic, she castigates those she takes to be the practitioners of "victim art" for representing "themselves to the public not as artists but as victims and martyrs" (1994/95, 54). The alleged focus of Croce's piece is a dance performance by HIV+ choreographer Bill T. Jones called "Still/Here." I say "alleged focus" because Croce refused to attend "Still/Here," but panned it anyway. In her "review" of "Still/Here," she accuses Jones of portraying victimization and suffering unmediated, because it includes, as background to the dance performance happening onstage, videotape of a variety of people discussing their experiences of various terminal illnesses, including AIDS and cancer. Croce knew about the video testimony because of the publicity surrounding the performance, which she notes "has been deafening" (54). The deafening publicity seems to have damned "Still/Here" in Croce's eyes, and it is based on this publicity, not the performance itself, that Croce claims she "understand[s] 'Still/Here' correctly" (54). That Croce's analysis is of an unseen object is pertinent; the piece becomes a projection of her own anxiety in the face of what she perceives as a danger to her role as critic in particular and to the purpose of art more generally. She believes a work like "Still/Here" leaves the critic no distance from the work, or from the life of its creator, and thus, no space in which to criticize or form a relation to the work that is independent of the work's relation to its creator's personal experience of suffering. "Still/Here," Croce is certain, is a "traveling medicine show," not art (54).[5]

I begin with the controversy surrounding Croce's analysis of "Still/Here" because it captures some key debates: about the relationship between illness as personal experience and sociocultural event, as well as the relationship between illness and art, philosophy, and politics at the end of the twentieth century and the beginning of the twenty-first century. Croce's piece also explicitly connects the emergence of "victim art" at the end of the twentieth century with the emergence of AIDS. She claims that dance and theater have changed (for the worse, in her view) because of AIDS, and because of the overtly political aspects of much art about AIDS. This is a disturbing phenomenon for Croce because what she wants is not the politicization of illness in and through art but the aestheticization of illness in the manner of the Romantics of the late nineteenth century. "The Romantics did not use art, they were used by it, consumed by it as much as by killer diseases," Croce writes. "The mass-produced art of the twentieth century, art that has no spiritual dimension, is art that you can use" (60). Art, according to Croce, should be disinterested, not useful, and, as such, concerned with the Spirit (with a capital "S") and not the body or even the self in the world. Croce's call for disinterested art, it seems to me, is a kind of nostalgia for the days when the explanatory categories of spirit/body, personal/political, and high art/low art were fiercely held apart. Many of the most compelling explorations of the experience of illness, including some of those in the case studies that follow, attempt to understand the way the distinctions between self, spirit, and body collapse in extreme and even everyday experiences of illness, and to arrive at conceptions of the self and spirit through the experience of embodiment, not in opposition to or distanced from it.

For Croce, AIDS is the enemy not because it is killing so many artists in the prime of their lives,[6] but because she believes work that deals with AIDS cannot be criticized. And so, because Jones's work *uses his own experience* of the death of his lover Arnie Zane of AIDS and his own experience of being HIV+, it cannot be criticized according to Croce. Tellingly, Croce blithely reduces Jones, as well as Robert Mapplethorpe, to "AIDS victims":

> Actually, I'm not sure who came first: Jones the AIDS victim or Mapplethorpe the AIDS victim. But Jones and Mapplethorpe, parallel self-declared cases of pathology in art, have effectively disarmed criticism. They're not so much above art as beyond it. The need for any further evaluation, formal or otherwise, has been discredited. Where will it go

> from here? If an artist paints a picture in his own blood, what does it matter if I think it's not a very good picture? If he mixes the blood with Day-Glo colors, who will criticize him? The artist is going to bleed to death, and that's it. (58–59)

For Croce, Jones and Mapplethorpe are AIDS victims and that's all; they are not artists, or artists of AIDS, or even, in a parlance meant to challenge such victimization, people with AIDS (PWAs). She seems to think that works about the experience of illness are simply and only attempts to portray the ill person as a victim or martyr to his or her illness, as opposed to works by Romantic artists, who are victims of and consumed by art.[7] In fact, Jones in "Still/Here" and many of the authors of the illness narratives I discuss try to do just the opposite; they attempt to portray themselves as *more than their illnesses*. Creativity in the face of death is most often a resistance to victimization and martyrdom, rather than, as Croce would have it, a giving in to it. Or, as Tony Kushner asserts in a letter in response to Croce's piece, "One of the chief gestures of good art about oppression is to question and refuse victim status, to exhort people toward agency and action" (1995, 11).[8]

Several commentators have found Croce's abdication of critical authority in the face of art that describes traumatic experience to be a frankly baffling way for a critic to respond; for these critics, Croce has thrown the baby of critique out with the bathwater of bad art. This is Frank Rich's main assessment of Croce's argument in a column for the *New York Times*. Like Sherwin Nuland, doctor and author of the book *How We Die* (1994), Rich attributes "the proliferation of art about disease and death to the AIDS epidemic," and agrees that "only some of that art enriches while much of it seeks only to 'take advantage of our empathy'" (1995, 19). Rich points out that "[l]ike the Holocaust or the Civil War, AIDS has been responsible for more kitsch than culture. For every poetic epic like *Angels in America,* there are bound to be at least a dozen examples of tear-jerking agitprop like the TV movie *And the Band Played On* (or worse). A critic's job is to separate the art from the junk rather than blur all distinctions out of ideological pique and dismiss the entire field" (19). I agree with Rich that there is a proliferation of art about disease and death partly because of the AIDS epidemic, though, as I note in chapter 2, the proliferation of writing and other art about the experience of the body generally began before AIDS, with the feminist movement of the late 1960s and early 1970s. And although I also agree that it is the critic's job to separate "poetic epic" from "tear-jerking agitprop," I also want to pay attention to the ways in which a particular culture constructs the boundary between poetic epic and tear-jerking agitprop, as well as the ways it determines the value of everything between and beyond these categories.

Is "Still/Here" poetic epic, tear-jerking agitprop, or something in between? Based on Croce's description of the performance in "Discussing the Undiscussable," one wouldn't have a clue what it is, though one may very well assume that there were no dancers on stage, only real live sick people talking about the experience (and I use the singular intentionally) of being ill. In Croce's piece there is little discussion of the process Jones used to make "Still/Here." Instead, she notes that she "can't review someone I feel sorry for or hopeless about," and one can't help but notice the slippage she performs in this sentence between reviewing a performance and reviewing a person.[9] She doesn't mention that Jones conducted "Survival Workshops" around the United States with people dealing with terminal illness, that these workshops were filmed, and that the testimonies recorded on film were edited and used as material for both the score of "Still/Here" and as visual backdrop and accompaniment to the dance. As the program notes, the lyrics for the "Still" section were "arranged by" Kenneth Frazelle, and the lyrics for the "Here" section were "arranged by" Vernon Reid, both of whom also composed the music, which was sung by Odetta. As Frazelle explains in the program, "I observed dozens of hours of videotapes of the Survival Workshops. As [visual artist] Gretchen [Bender] and Bill [T. Jones] and I decided what the song topics might be, specific voices and lines from the workshop participants converged into song texts."[10]

The testimonies gathered at the "Survival Workshops" are not "Still/Here" in any simple, unmediated fashion; they are a part of the texture of the performance about experiences of illness. Or, as Jones himself explained, the "Survival Workshops" were "scientific research" (Walker Art Center 1998, 130). Although Jones acknowledges that his own personal experience with HIV and AIDS contributed to his decision to make "Still/Here," the final product is in no way his personal experience of AIDS unmediated.[11] In a conversation with Thelma Golden, Jones describes how he came to conceive of the piece:

> I looked to the personal. Concern about my own HIV status, concern about what AIDS has meant. Also wanting to redefine this phenomenon as part of a continuous and cogent fabric. Therefore I decided to do scientific research. Rather than mine my emotions, to go connect with and question different persons living in contrasting situations. I would gather material and make a theater work from it. It's not an original idea, but it is what I had to do. So I thought it wouldn't be at all controversial. I thought it might be illuminating,

> I would learn a lot, and I thought other people might also if the work was conceived well. We would be able to come closer to these questions and own them, especially people who were not ill. That is what I was doing. (Walker Art Center 1998, 130)

The piece was crafted—"It was costumed, lit, choreographed, sung. It was a show" (130); yet it also sought to "push the boundaries of art and life as far as you can. Mix them up" (131). "Still/Here" is not unlike many of the illness narratives that I look at in the chapters that follow: while it seeks to break down some of the boundaries between the person who is ill and his or her witnesses, it also acknowledges that suffering represented in art is *never* unmediated and that attempts to communicate suffering from one person to the next often fail.

I begin *Treatments* with Croce and Bill T. Jones not because this book is about performing art, though it is about the performance of the self in one particular art form: memoir. Rather, I begin with Croce and Bill T. Jones because I want to challenge not only Croce's notion of "victim art" but also her understanding of "experience." To believe, as Croce seems to, that "Still/Here" is *personal experience unmediated* is to deny the gap between experience itself (whatever that may be) and representations of experience. In her essay on the use of experience as historical evidence, feminist historian Joan Scott challenges the notion of "writing as reproduction, transmission[, or] the communication of knowledge gained through (visual, visceral) experience" (1992, 24). Rather than relying on "experience as the origin of knowledge" of the individual subject, Scott prefers to emphasize "the constructed nature of experience" (25). She is interested, therefore, in the "categories of representation," and how these categories operate within given ideological systems. "Experience" itself is an explanatory category, Scott asserts, and reinforces her point by placing the word in quotes (26).

It is interesting in relation to my project that Scott begins and ends her discussion of "experience" as historical evidence with readings of Samuel Delany's memoir, *The Motion of Light in Water,* about being homosexual in New York in the early 1960s. According to Scott, Delany's memoir is evidence of alternative modes of being (homosexual) and alternative spaces (for homosexual sex and community) that more conventional history fails to acknowledge or even see. Scott uses Delany's memoir as an example of a text that "drastically raises the problem of writing the history of difference, the history, that is, of the designation of 'other,' of the attribution of characteristics that distinguish categories of people from some presumed (and usually unstated) norm" (22). What is important about the rendering of "experience," according to Scott, is not simply to make visible that "experience," but rather to reveal the ways that "experience" is not a reliable or self-evident source of knowledge. The category of "experience," in other words, as Meaghan Morris has claimed, might be used more effectively "as a way of posing skeptical questions of history" (1998, xxii), and memoir as a necessarily aestheticized *and* politicized record of "experience" often participates in this skeptical questioning.

The main argument against memoirs in general is much like Croce's critique of "victim art." According to this argument, *all* memoirs are nothing more than a self-indulgent mining of personal experience for more and more extreme events that the author hopes will engender sympathy, disgust, or both from his or her readers. Because memoir, unlike fiction, supposedly describes "true" experience, it can't be criticized as "art," as Croce might contend. Such a position reflects the ongoing backlash against memoir, signaled by an increasingly common critical response to memoir as something other than "literature." In 1996, before this backlash, the *New York Times Magazine* published a special issue entitled "True Confessions: The Age of the Literary Memoir," which included a critical piece by James Atlas and "real-life stories" from, among others, Susan Cheever, Chang-rae Lee, Phyllis Rose, Luc Sante, and Art Spiegelman. Though it voiced many of the current concerns with the memoir boom ("the license to tell all has produced a virtual library of dysfunctional revelation") and stressed that the "habit of self-examination can grow tiresome," Atlas's essay also emphatically stated that "[f]iction isn't delivering the news. Memoir is. At its best, in the hands of a writer able to command the tools of the novelist—character, scene, plot—the memoir can achieve unmatchable depth and resonance" (26).

Just three years later, in an essay published in the *Chronicle of Higher Education* entitled "Why Memoir Isn't Always Art," Brock Clarke discusses the recent popularity of memoir "and why we should be wary of it" (1999, B9). Much more subtly than Croce, Clarke makes the distinction between confession and creative art, asserting that the "simple narration of one's life is not necessarily art" (B9).[12] Clarke agrees with the preeminent complaint against memoir, that it is self-indulgent, that it is, as Clarke puts it, "concerned less with the relationship between the self and the world, and more with the self *as* the world" (B9). But his grudge against the rise of the memoir has more to do with its supposed simplicity than with its self-indulgence.[13] Clarke quotes the novelist, critic, and biographer Jay Parini, from an article also published in the *Chronicle* (1998), who asserts that his students prefer memoir to fiction because, through reading memoir, they *more easily learn things* than through reading fiction. Clarke's very necessary question is, "should our access to art be so easy?" Because life is difficult, according to Clarke, "we need to find ways to represent its difficulty; and we represent life's difficulty by adding to it, by putting it into a form shaped primarily by aesthetic sensibility, not by our own experiences" (B9). Like Croce, therefore, Clarke seems to think experience can be rendered unmediated and is different from an aesthetic sensibility; and so he comes out in favor of fiction (as Croce comes out in favor of Romanticism), as if fiction is always good and memoir is always bad, and as if fiction is only imaginative and not true and memoir only true and not imaginative.

In *How Our Lives Become Stories* (1999), Paul John Eakin considers contemporary memoirs and the ways in which they reveal the processes of "making selves" (x) and the essentially "embodied nature of selfhood" (1). In his final chapter entitled "The Unseemly Profession," Eakin looks at the ethics of memoir writing, and the question of what drives people to tell the stories they tell. In order to delve into these ethical questions, Eakin takes the case of Kathryn Harrison's controversial memoir, *The Kiss* (1997), about the incestuous relationship she had as an adult with her father, and considers both the work itself and the public and critical response to it. As Eakin shows, there has been much critical soul searching about what has been called "the age of memoir." The possibility that we are living in an "age of memoir" not only says much about the literary and publishing industry, but also, more generally, it says much about the way we live now. Eakin sifts through the impassioned outpouring in response to Harrison's book, delineating the arguments of her critics who suggest, for example, "that Harrison exploited the memoir genre to give her sales a boost" (151, note 7) and that such work is little more than "merchandising pain" (153), as well as the arguments of her supporters who believe Harrison "answered to an imperative of the creative imagination, an 'artistic' necessity overriding any other consideration" (152–53), and that Harrison's prodigious talent (which, remarkably, most reviewers agree on) places her "in the company of some of [memoir's] most admired twentieth-century practitioners, including George Orwell, Mary McCarthy, Frank Conroy, and Mary Karr" (155).

Throughout *Treatments*, I will touch on the position of memoir in culture and, like Eakin, will take up the question of ethics with regard to this particular genre. But, for now, I want simply to make a rather obvious point: there is good memoir and bad memoir (and everything in between). Even Clarke concedes this point when he admits that not all contemporary memoirs "legitimize qualities—sentimentality, caricature, vitriol—that we should abhor in art" (1999, B9).[14] The rise of memoir is not inevitably a sign of the demise of literature. Those who see memoir as a degeneration of literature are, like Croce, nostalgic for a (literary) past that never was, and guilty of the sort of sweeping criticism that would be unimaginable if made of another genre, of, say, the novel or drama.[15] Such overarching criticism makes its way almost daily into reviews of new memoirs. Thus, in a glowing review of Jerome Charyn's memoir *The Black Swan*, Leslie Epstein opens with an example of what has become characteristic critical dismay over the entire genre:

> Anyone looking back over the last century can mark the progressive enfeeblement of the literary imagination by comparing the best novels of its first half, incontestably *Ulysses* and *Remembrance of Things Past*, with those of its second—perhaps *Midnight's Children*, perhaps *Lolita*. The nadir of the decline seems to be occurring at the century's turn, with the concomitant rise of memoir. Out of the contemporary parade of alcoholic mothers, abusive fathers, over- and under-eating and flesh that has been mutilated, mortified and indulged, Jerome Charyn's *Black Swan* steps forth as a robust and instructive exception. (Epstein 2000, 13)

"Progressive enfeeblement of the literary imagination" and the "nadir of the decline" recalls the rhetoric associated with the eugenics movement in the first half of the twentieth century. In this not uncommon assessment, memoir is a degenerate genre, lacking virility and robustness, and its rise indicates the decline of culture signified by its feminization and personalization. Memoir, it seems, is the sick woman (or, in the age of AIDS, perhaps the sick gay man) of literature. Nancy Mairs, who has written memoirs that deal with everything from her experience of multiple sclerosis to her husband's cancer and impotence, from both of their extra-marital affairs to her relationship with Catholicism, has described much contemporary memoir as "the literature of personal disaster" (1994) and has noted "that more women than men write, and write more intimately, about physical and emotional distress" (1997, xi). The backlash against memoir has, therefore, much, if not everything, to do with literary gender politics, as well as imagined national selves as exemplified in popular national literary forms, a point I will discuss at greater length in chapter 3. As I show throughout *Treatments*, the particular form a discourse about illness takes matters: not just in the obvious way that different forms reveal experiences of illness from different perspectives, but in a more constitutive way. The ways in which we are able to think and do illness come into being through the narratives and practices that already structure the experience and event of illness. *Treatments* explores the discourses and practices already available for speaking and doing illness, who is considered an expert of these discourses and practices, and how new discourses and practices of illness come into being.

But to reiterate, to judge a memoir is not to judge a life but to judge a representation, which is always partial and contingent, and determined as much by the reader and what she brings to the text as by the author. This is not to say, however, that attempts to render experience in language are misguided. Inadequate? Inevitably. Misguided? Sometimes, perhaps, but not always. In *Treatments* I hope to discern why memoirs in general and illness narratives in particular are often inadequate and yet still necessary.

Effective and Affective Histories: The Case Studies

My approach to illness narratives is less concerned with evaluating the literary merit and more concerned with what this sort of work does personally, politically, and culturally. To investigate what illness narratives do, I consider the ways that illness narratives describe both *particular, individual experiences of illness*, as well as the ways that illnesses, such as tuberculosis, AIDS, and breast cancer, become *cultural and political events*. In order to clarify what I mean by illness experiences and illness events, I need to first delineate how I conceive of illness narratives as both affective and effective histories.

"Effective history," according to Foucault in a discussion of the term he takes from Nietzsche, is opposed to any notion of historical objectivity and, in turn, is concerned with doing historical work that has political meaning and utility for the present (Foucault 1980, 64). In an interview published in the collection *Power/Knowledge*, Foucault states:

> [I]f one is interested in doing historical work that has political meaning, utility and effectiveness, then this is possible only if one has some kind of involvement with the struggles taking place in the area in question. I tried first to do a genealogy of psychiatry because I had had a certain amount of practical experience in psychiatric hospitals and was aware of the combats, the lines of force, tensions and points of collision which existed there. My historical work was undertaken only as a function of those conflicts. The problem and the stake there was the possibility of a discourse which would be both true and strategically effective, the possibility of a historical truth which could have a political effect. (64)

For someone whose work is often accused of evacuating the political, the above comment situates political efficacy (if not precisely agency) at the very heart of Foucault's own discursive and nondiscursive practices. What is at stake for Foucault is nothing short of a "historical truth which could have a political effect." Foucault's idea of history, then, "disconcerts"[16]—that is, it opposes the "consoling play of recognitions," even, or especially, at the level of the individual body, which, Foucault notes, is never "sufficiently stable to serve as the basis for self-recognition or understanding" (1977b, 153). Foucault suggests genealogy as a historical method that seeks to unravel the progressive and linear history of the historicists.[17] While the historicist examines and instantiates origins, the genealogist traces lines of descent in all their complexities. The genealogist looks for discontinuities, but also, paradoxically, surprising continuities, those echoes of counternarratives that reverberate across time and space, waiting for the future to hear them. Although I return again and again to Foucault's historical work—on the emergence of the anatomo-clinical method described in *The Birth of the Clinic*, on the operation of biopower discussed in *Discipline and Punish* and in his lectures at the Collége de France, on the "arts of existence" and forms of subjectification formulated in his three-volume *History of Sexuality*—I also demonstrate that his work sometimes fails to provide a theoretical or methodological model for reading experiences of loss such as those articulated in illness narratives. And so, while Foucault's notion of *effective* history is necessary in order to demonstrate discontinuity and rupture at the sociocultural level (though as always inscribed on individual bodies), illness narratives also demonstrate discontinuity and rupture at the subjective and intersubjective level. It is necessary, therefore, to do affective as well as effective history in order to grasp the ways in which the breakdown of the body that occurs in illness provides a phenomenological reduction of sorts that allows one to look anew upon that which one has formerly taken for granted, in terms of one's relationship to the world, to others, and to the self.

By doing both effective and affective history I approach the figure of the ill person and the writings that emerge out of the experience of illness from, in the terminology of Elizabeth Grosz, both the "outside in" and the "inside out."[18] The return of the repressed body in twentieth-century philosophy is connected, I think, to a similar return of the body in the late twentieth-century literature that is the object of my analysis. In *Treatments*, therefore, I utilize the work of several twentieth-century thinkers, including Foucault, Merleau-Ponty, Lacan, Gilles Deleuze, Félix Guattari, Elaine Scarry, Eve Sedgwick, Maurice Blanchot, and Jean-François Lyotard, who have placed the body at the center of their thought and who, read with and against each other, suggest affective and effective approaches to thinking about and through the body. I hope to move between effective and affective theoretical modes, to oscillate from the outside in to the inside out, because in the case of illness narratives there is always both a movement in (the embodied self in relation to itself and to death) and a movement out (the embodied self in relation to others; to institutions, including, in particular, the institution of medicine; and to communities, national and otherwise); not either/or but both/and, an oscillation unto death and, within narrative, beyond death.

In the chapters that follow, I present case studies of how one might do treatments of particular illness narratives. By comparing two or more illness narratives in each chapter, I hope to present illnesses as experiences that individuals not only have but do, and as sociocultural and historical events. My first case study looks at the experience and event of tuberculosis in the United States in the mid-twentieth century, before the widespread use of antibiotics that, for a time at least in the United States, successfully countered TB. In this chapter, I delineate my use of several of Foucault's key concepts, including the "anatomo-clinical method," "biopower," and "subjugated knowledges," and offer treatments of Betty MacDonald's memoir *The Plague and I* (1948) and *Madonna Swan: A Lakota Woman's Story* (published in 1991 in a volume edited by Mark St. Pierre, but describing Madonna Swan's experience of TB in the 1940s and early 1950s). I intend this discussion to set the historical stage for a shift that I then describe in chapter 2.

After juxtaposing Foucault's sociohistorical articulation and MacDonald's and Swan's literary representations of the ways in which biopower operates, I next introduce the possibility that at the end of the twentieth century we entered a new episteme, one that was characterized by the emergence of a new figure[19]—the politicized patient—and a new genre—the patient's counternarrative to medical discourse as exemplified by doctor's charts and case histories.[20] In order to suggest the complexity of this moment of emergence, I discuss three writer-activists—Susan Sontag, Audre Lorde, and Eve Kosofsky Sedgwick—and the particular forms of politicization their work demonstrates. Through readings of Sontag's *Illness as Metaphor* (1978) and *AIDS and Its Metaphors* (1988), Lorde's *Cancer Journals* (1980), and Sedgwick's "White Glasses" (first presented in 1991 and published in 1993 in *Tendencies*), I investigate how each author does illness and understands politicization. Although all three have been outspoken breast cancer survivors and were motivated to write their books because of their own experiences with cancer, Sontag chooses, for reasons I will delineate in chapter 2, not to tell an affective history, while Lorde and Sedgwick, on the other hand, tell illness narratives that are both effective and affective histories.

I situate the emergence of this new figure and genre in 1980, but not because *The Cancer Journals* is the first counternarrative to medical discourse from the patient's perspective. As I discuss, there are many earlier patients' narratives, and some of them might even be read as offering politicized—or proto-politicized—modes of being ill and doing illness.[21] What is significant about this historical moment, however, is that it *draws on* feminist health knowledges and practices that emerged in the 1970s and *anticipates* the knowledges and practices that would emerge in the wake of the AIDS crisis in the 1980s. The politicized patient that emerged out of the feminist health and AIDS movements would be, like Sontag, Lorde, and Sedgwick, both activist and writer and, as such, determined to transform society and its institutions, especially medicine, as much as, if not more than, the self who is ill.

Through reading countless illness narratives published in the past twenty years, I have discovered that while some illness narratives offer a counternarrative that articulates the subjugated knowledge of the patient, many others offer narratives that articulate hegemonic medical knowledge. Thus, an important question throughout *Treatments* will be: How are subjugated knowledges covered over? How are counternarratives appropriated by and made appropriate for dominant institutional knowledges? I contend that by the turn of the twenty-first century, calls for the political transformation of society in general and the institution of medicine in particular are drowned out by more fervent cries for the personal transformation of the self. I end chapter 2 with an analysis of Barbara Ehrenreich's widely read essay in *Harper's*, "Welcome to Cancerland" (2001), in which she deplores this depoliticization—or perhaps neoliberalization is a better word—of patienthood, and critiques what she calls "breast-cancer culture" and its intimate relationship with, rather than challenge to, the "cancer industrial complex."

After contrasting two competing discourses of patienthood—politicization and personalization—in chapter 2, I offer another comparison in chapter 3, this time between what I suggest are national arts of being ill and doing illness. I continue to use Foucault's methodology to diagram and diagnose an American art of being ill, which emphasizes the cultivation of an improved self, in contrast to a British art of being ill, which emphasizes the cultivation of an ironic self. In order to discuss the American versus the British arts of being ill, I take up Sandra Butler and Barbara Rosenblum's *Cancer in Two Voices* (1991) and Ruth Picardie's *Before I Say Goodbye* (1998) as exemplary cases of different national arts of being ill. In doing so, I do not mean to claim that there is only one British and one American art of being ill, nor do I believe that what constitutes a nation or a national identity is either uncontested or ahistorical. What I do mean to suggest, however, is that people who write about the experience of illness often draw upon certain cultural conceptions of the nation and its institutions. The national arts of being ill that I attempt to delineate are fictions, but my hope is that to delineate these fictions will be useful in better understanding the specific cultural narratives of illness that are available in different sociocultural contexts, as well as the particular cultural anxieties that these narratives often keep concealed.

My fourth chapter treats illness narratives written by doctors rather than patients. In this case study of physicians' narratives I consider the possibility that doctors themselves can disrupt the hierarchical doctor-patient binary through their own experience of patienthood. Such is the case in the memoir of Oliver Sacks (1993), in which he describes a leg injury he sustained that resulted in neurological damage. It is also evident in the potentially revolutionary process of what I call *becoming-patient*, as exemplified in the memoirs of two doctor-writers—Abraham Verghese and Rafael Campo—who treated people with AIDS in the 1980s and early 1990s. My concept of becoming-patient draws on Deleuze and Guattari's articulation of the possibility of nonteleological, minoritarian becomings, which are movements away from unitary, universal, and stable identifications toward not some other unitary, universal, and stable identification but toward other becomings.

As in my chapter on the politicization of patienthood, in this chapter on doctors' narratives I show the possibility of resisting the power/knowledge nexus that Foucault called "biopower," this time from the doctor's side of the binary. Deleuze and Guattari are interested not only in the ways in which power/knowledge operates but also in the minority discourses (like Foucault's "subjugated knowledges") that can and do emerge not from beyond or behind the operations of power/knowledge, but from within it. This does not mean, however, that doctors become patients in reality, but that they are constantly becoming-patient through the processes of *desubjectification* (a giving up of their identity as, or only as, doctor within the doctor-patient binary), *deterritorialization* (a movement away from the institutional spaces and practices of medicine that create distance rather than proximity between doctor and patient), and *disarticulation* (an unlearning of certain modern medical knowledges and practices cultivated in medical school). I consider the possibility of doctors' becoming-infected and becoming-affected by their patients; again, not in reality, by HIV itself, but by their patients' stories.

As enthusiastic as they are of the possibilities of becoming-other, and as much as they themselves attempt to enact such becomings in their own writing, Deleuze and Guattari are, nonetheless, aware that most often such becomings fail. According to their theories of becoming, we cannot help but resubjectify, reterritorialize, and rearticulate the socially and institutionally accepted forms of being. In the second half of chapter 4, therefore, I consider the ways in which, in the case of HIV/AIDS in particular, some doctors' narratives not only enact the possibility of *an-other medicine*, in which doctors are infected and affected by their patients, but also inscribe, or reinscribe, the reality of the same old medicine, in which doctors are omnipotent and omniscient figures and patients are the objects on which the doctor's power/knowledge is enacted.

In chapter 5, I consider what I call the practices of witnessing that emerge in two writers who describe the experience of their partners' dying and their own interminable mourning: Paul Monette's autobiographical writings on the death of his lover Roger Horwitz from AIDS and John Bayley's work on the death of his wife, the philosopher and novelist Iris Murdoch, from Alzheimer's disease. Monette's and Bayley's narratives of dying present situations of being, as Lacan describes it in his ethics seminar, between two deaths: that is, between a symbolic, social death and a real, absolute death. In the case of Monette, the death of Roger is the death of the other that is also the symbolic death of the self. Monette, in a manner that will exemplify yet again the American art of being ill that I describe in chapter 3 and touch on again in chapter 4, turns to elegy and classical heroism as the narrative condition for the possibility of compensation for Roger's (and eventually his own) death. In the case of Bayley, Murdoch's Alzheimer's is a kind of symbolic death of the self before the real death of the body. In contrast to Monette, Bayley exemplifies the British art of being ill with his presentation of an anti-elegy in which humor—and, indeed, *nonsense*—is the condition for the possibility of compensation for Murdoch's lost self. Both authors cultivate a "response-ability," in the terminology of feminist philosopher Kelly Oliver, to their partners as they become overwhelmed by AIDS and Alzheimer's, and to others beyond them, who are also overwhelmed by AIDS and Alzheimer's.

In my final case study, I describe explicitly the possibility of a particular ethics, which I call an ethics of failure, that I think emerges implicitly out of the cases described in the preceding chapters. In order to explain what I mean by an ethics of failure, I turn to Gillian Rose's illness narrative, *Love's Work* (1995), which relates her experience of ovarian cancer and her ethical formulation "being at a loss yet exploring various routes," and Atul Gawande's collection of essays entitled *Complications: A Surgeon's Notes on an Imperfect Science* (2002). Both of these texts seek new routes or new idioms across the gap between doctor and patient, but from different positions in the doctor-patient binary. I understand this gap as, in Lyotard's terminology, a *différend*, or a fundamental incommensurability between the multiple idioms within medicine. What aesthetic, ethical, and political practices must we invent to communicate across this incommensurability? I concede that an ethics of failure at first glance is hardly inspiring, as a reader of an earlier draft of *Treatments* noted. That reader suggested, I think correctly, that through my concept of an ethics of failure, I am in a sense also describing an ethics of transformation, risk, and possibility. Nonetheless, in *Treatments* I want to revalue not only loss, but also failure: the idea that we are fallible, that we get things wrong, that we might not be able to do, know, or feel enough, or that we might do, know, and feel too much.

It bears mentioning at this juncture that Foucault would perhaps express some methodological reservations about my project, if we heed his description of his archaeological method in *The Archaeology of Knowledge* (1972). Foucault, whose own work rarely dealt directly with the present, believed that it is impossible to describe one's own "archive," or "the general system of the formation and transformation of statements" (130), precisely because one is within it, and not outside or beyond it.[22] The archive, in other words, is that which delimits what one can and cannot say; there is no thought from outside one's own archive. And yet, Foucault also submits in the introduction to *The Archaeology of Knowledge*, "I have tried to define this blank space from which I speak, and which is slowly taking shape in a discourse that I still feel to be so precarious and so unsure" (17). For Foucault it is a blank space *from* which one speaks, not a blank space *of* which one speaks. And it is this blank space that connects Foucault, the archaeologist of knowledge and theorist of power, with Maurice Blanchot, the writer of fiction and theorist of the literature of disaster, whose work Foucault discusses in his essay "Maurice Blanchot: The Thought from Outside" (1986b). I juxtapose Foucault and Blanchot in order to signal how we might bring the effective and affective together, and to suggest why it is necessary to do so.

In his archaeological and genealogical work, Foucault describes discursive and nondiscursive formations, which together form an apparatus "always inscribed in the play of power" (1980, 196). Can one imagine and speak of something outside of these relations of power/knowledge? No, according to Foucault. And yet, we imagine (new worlds). And yet, we speak (new words). And yet, we create (new forms). And yet, we listen (to the echoes of other voices, other knowledges). Where Foucault provides us with a mapping—an architectural rendering of sorts—of this power/knowledge nexus, Blanchot offers us a reckoning in words of what words cannot say: "When all is said, what remains to be said is the disaster. *Ruin of words, demise of writing, faintness faintly murmuring: what remains without remains* (the fragmentary)"

(Blanchot 1995, 33; emphasis in original). What connects Foucault and Blanchot, and what connects both to my project, is not only the difficulty of effective agency in the face of power (Foucault) or of true and meaningful words in the face of suffering (Blanchot), but that these moments of difficulty are always also generative. *Treatments* seeks to negotiate with what might emerge out of failure, the creative possibilities arising out of those difficulties the individual encounters within an archive of treatments that are already there for the person who is ill, though these treatments are never complete or final. In the case studies that follow, I do treatments of the multiple ways illness is done, in and through the body, narrative, and politics.[23]

1.

Patients and Biopower

Disciplined Bodies, Regularized Populations, and Subjugated Knowledges

"This is a book about space, about language, and about death; it is about the art of seeing, the gaze" (ix). So begins *The Birth of the Clinic* (1973), Foucault's "archeology of medical perception" that describes the emergence at the end of the eighteenth century of what he calls the "anatomo-clinical method" (4). In looking back, first at Foucault's treatment of the emergence of a particular method of medical perception, and then at two mid-twentieth-century memoirs that are also about space, language, and death with regards to the experience of tuberculosis in the mid-twentieth-century United States, I hope to provide a historical and theoretical framework through which to read late-twentieth and early-twenty-first-century illness narratives. The narratives—Betty MacDonald's memoir *The Plague and I* and Madonna Swan's oral history *Madonna Swan: A Lakota Woman's Story*—are exemplary cases that illuminate the spaces, practices, and objects of tuberculosis prevention and treatment in particular and the operations of biopower more generally, which Foucault describes as the "numerous and diverse techniques for achieving the subjugation of bodies and the control of populations" (1978, 140).

On the one hand *The Plague and I* and *Madonna Swan: A Lakota Woman's Story* demonstrate the anatomo-clinical method and the two modes of biopower: the disciplining of bodies and the regularization of populations. On the other hand, these illness narratives might also be understood as articulating "subjugated knowledges," which Foucault defines as low-ranking or naive knowledges that remain marginal to institutional knowledges, including the institutional knowledges of medicine. For Foucault, the investigation into subjugated knowledges in the past opens up a space for thinking, being, and doing otherwise in the present and future. He asks that we actively seek these knowledges in the past because

> it is through the re-emergence of these low-ranking knowledges, these unqualified, even directly disqualified knowledges . . . and which involve what I would call a popular knowledge *(le savoir des gens)* though it is far from being a general commonsense knowledge, but is on the contrary a particular, local, regional knowledge, a differential knowledge incapable of unanimity and which owes its force only to the harshness with which it is opposed by everything surrounding it—that it is through the re-appearance of this knowledge, of these local popular knowledges, these disqualified knowledges, that criticism performs its work. (1980, 82)

Subjugated knowledges always threaten to disrupt the power/knowledge nexus; they are something of a Freudian "return of the repressed" in terms of socio-cultural rather than individual histories. Relatedly, historian Michel de Certeau reads Freud's *Civilization and Its Discontents* (1961 [1930]) as in part discussing the "overturning moment" in postmodernity from specialization to general knowledge, the knowledge of the "ordinary man" (and woman), similar to Foucault's popular knowledge *(le savoir des gens)*. The ordinary man or woman in the twentieth century becomes a narrator, and I am concerned with the ways that the ordinary insinuates itself into established scientific fields, including medicine, through popular genres, including illness narratives (Certeau 1984, 4–5).

This "overturning moment" of postmodernity is also discussed in Arthur W. Frank's important work, *The Wounded Storyteller* (1995), in which he suggests, as I do, that the contemporary proliferation of illness narratives from the patient's perspective is a necessary counternarrative to the narratives of doctors and other medical professionals. Frank's work serves as something of a model for mine, in that he grapples with literary critical as well as philosophical and sociocultural issues in relation to illness narratives.[1] He identifies the figure of the "wounded storyteller" as having emerged in postmodern times, because in postmodernity, "the capacity for telling one's own story is reclaimed" (7).[2] While Frank is hopeful about what this emergent capacity signifies, he is not naïve about its transformative effects; he knows that the medical narrative will not—indeed, should not—be drowned out by the voices of patients, and, moreover, that the "postmodern claim to one's own voice is halting, self-doubting, and often inarticulate" (7). Nonetheless, despite, or perhaps because of, the inarticulateness of such voices, the emergence of these stories is transforming how illness in particular, and the body in general, is experienced and written.

According to Frank, the modern experience of illness is substantively different from the postmodern experience of illness.[3] Frank identifies Talcott Parsons as the preeminent theorist of the modern experience of illness, an experience characterized by the patient's assumption of what Parsons identified as the "sick role," which required an absolute surrender to the care of a physician and the institutional structures of medicine. Partly because of the shift in the twentieth century from acute to chronic illnesses (thanks, of course, in no small part to the successes of modern medicine), the experience of illness has shifted as well. Formerly patients generally either got better or died; now, however, patients are often neither fully cured of nor dead from their illnesses, but, rather, are in and out of remission. The body remains haunted by illness and its threat; the person is neither fully healthy nor precisely ill; they are somewhere in between. Frank maintains that in postmodern times there are many people, including himself,[4] who occupy this liminal position between health and illness: thus, in the late-twentieth-century West, we live in what Frank calls a "remission society" (8; see also Wendell 1996). In a "remission society" the boundaries between health and illness are permanently disrupted, thereby challenging the dichotomous formulation of health as the norm and illness as that which deviates from the norm. In her book on the representation of disability in literature and culture entitled *Extraordinary Bodies*, Rosemarie Garland Thomson similarly notes that because anyone can become disabled (through age, injury, or chronic illness) at any time, disability is "more fluid, and perhaps more threatening, to those who identify themselves as normates than such seemingly more stable marginal identities as femaleness, blackness, or nondominant ethnic identities" (1997, 14).

As Frank understands it, the postmodern does not in any way mean an evacuation of the ethical, as many commentators have suggested.[5] Neither does the proliferation of illness narratives necessarily signal a relativization of experience; rather it potentially represents no less than a *radical democratization of medicine*, which relates again to what de Certeau diagnoses as the insinuation of the ordinary into the scientific. "Telling stories of illness is the attempt, instigated by the body's disease," Frank writes, "to give a voice to an experience that medicine cannot describe." He continues:

> The voice of the ill person is made possible by modernist medicine, but it cannot be contained within modernist assumptions, particularly those about medical professional dominance and the narrative surrender this dominance requires. A divide has been crossed into new territory, the post-modern, and we know this crossing by the new voices that are heard. (1995, 18)

For Frank, then, telling stories—and telling stories *of* and *through* the body in particular—is, simply put, an ethical act, with the potential to "*repair* the damage that illness has done" (53, emphasis in original). But in many contemporary illness narratives, to repair is not the same as to return to the way one was before. The repaired self always remains haunted by past and potential future experiences of illness. Frank associates illness narratives with other forms of what he calls "self-stories," which include stories of the self in relation to spirituality, gender identity, and trauma, and he believes that one of the reasons for the proliferation of these self-stories now has to do with "the accumulated violence of modernity," which is enacted upon real bodies and selves, and the stories they tell.[6]

In *The Plague and I* and *Madonna Swan*, we are presented with the subtle and not-so-subtle violence of modernity as it is enacted on real bodies, selves, and entire populations. Although both MacDonald and Swan are caught in the structures that Foucault and Frank delineate in their work, they also challenge those structures, in and through their particular narratives of illness. Through the case studies in this chapter and the next, I attempt to discern shifts in patients' narratives about the experience of illness from the mid-twentieth century to around 1980, a moment when I suggest a figure called the politicized patient emerges. In *The Plague and I* and *Madonna Swan*, MacDonald and Swan are not yet politicized patients, and doctors are still heroic figures whose expert knowledge gives them power over their patients. Nonetheless, through the humor that punctuates MacDonald's story and the ghosts that haunt Swan's, the subjugated knowledges of these two TB patients might still be read as counternarratives to the hegemonic narratives of medicine.

Foucault's Speaking Eye

In order to understand the mechanisms of the anatomo-clinical method and its relation to the operation of biopower, I first investigate the key terms Foucault identifies in the opening sentence of *The Birth of the Clinic* (1973): space, language, and death. The space to which Foucault refers is both an external space and an internal space: the space of the clinic and the space of body. The external space of the clinic is the "meeting place of doctor and patient" (111), a domain in which the techniques of interrogation and examination take place. Doctor and patient do not meet on equal terms, however; the clinical situation is "non-reciprocal" (xv), as the terms "interrogation" and "examination" imply. In order to determine the space of disease, to precisely locate the "pathological fact"[7] within the patient's body, "the doctor must abstract" the patient. "Paradoxically, in relation to that which he is suffering from," Foucault notes, "the patient is only an external fact; the medical reading must take him into account only to place him in parentheses" (8). In this procedure, the patient is individualized, and yet still objectified; that is, she is individualized as a body, not as the subject of her own experience of illness.

Although it gives "suzerainty" to the gaze, the anatomo-clinical method, as Foucault's second term in his opening sentence indicates, makes possible the determination of a pathological fact through language as well as, or in correlation with, the gaze. According to Foucault, what is essential in this method is an alteration between examination and interrogation, between "perceived stages" and "spoken stages" (111). Language and the gaze converge on the object: the pathological fact extricated from the patient's body. The ideal image of such a convergence between language and gaze is that of a "speaking eye" (114) that provides an "exhaustive description" (113). "Description, in clinical medicine," Foucault explains, "does not mean placing the hidden or the invisible within reach of those who have no direct access to them; what it means is to give speech to that which everyone sees without seeing—a speech that can be understood only by those initiated into true speech" (115). In this domain, the patient is not capable of true speech, even though she may speak of her illness; rather, she is the ground upon which this initiation—the doctor's initiation—takes place. Like the disease itself, the patient's words are objects to be interpreted by the doctor, to be translated by the doctor's "speaking eye" into a pathological fact, and incorporated into a case history.

The anatomo-clinical method makes the body speak—"all that is *visible* is *expressible*" (115; emphasis in original). However, it is not the patient's voice that matters, but the doctor's voice offering an exhaustive description of the patient's abstracted body. This becomes clear when Foucault brings in his third term—death—in his analysis of the anatomo-clinical practice of pathological anatomy. Foucault quotes Bichat's *Anatomie générale* of 1801: "Open up a few corpses: you will dissipate at once the darkness that observation alone could not dissipate" (146). With pathological anatomy, the internal spaces of the body are revealed; in this scene of literal and figurative illumination, "the absolute eye [of knowledge] that cadaverizes life and rediscovers in the corpse the frail, broken nervure of life" (166). Death, paradoxically, makes visible that which life had kept in the shadows.[8]

Foucault places this transformation of medicine—this emergence of the anatomo-clinical method—in the very midst of a transformation of the "philosophical status of man" (198). Medicine becomes one of the "sciences of man," contributing to the methodological and ontological possibility for the individual to "be both subject and object of his own knowledge" (197). Thus, Foucault ends *The Birth of the Clinic* by linking medical thought with philosophical thought and "lyrical experience," that is, with the connection between "figures of knowledge" and figures in language (198). The dominant medical thought, philosophical thought, and literature of an episteme all reveal "the fundamental structures of experience," an experience that encompasses not only life, but "forms of finitude," including, of course, death.

Tuberculosis as Experience and Event

In their narratives, both MacDonald and Swan discuss diagnosis, treatment, and sanatorium life. Although both women are stigmatized as a result of their diagnosis, their sanatorium experiences are distinct in many respects, reflecting the difference between the *event of tuberculosis* in a mostly white community and in a poor Native American community. By focusing my discussion on the subjugation of individual bodies as described in MacDonald's text and on the control of populations, and the racism that justifies such control, as described in Swan's text, I do not mean to imply that MacDonald's work does not also demonstrate the control of populations nor that Swan's work does not also demonstrate the subjugation of bodies. Indeed, the subjugation of bodies and the control of populations are related operations of biopower that create overlapping domains of intelligibility at work in institutions, such as public health and medicine, as well as in narrative forms, including memoir and oral history.

In her groundbreaking work *Reconstructing Illness* (1993), Anne Hunsaker Hawkins uses the term "pathography"[9] for illness narratives, and she defines pathography as "our modern detective story," where we are transported out of the everyday, familiar world of health into the unknown, uncharted world of illness (1). Hawkins asserts that, as a genre, pathography "seems to have emerged *ex nihilo*; book-length personal accounts of illness are uncommon before 1950 and rarely found before 1900" (3). She wonders why this is so, and offers as a possible reason the fact that pathography seems to have replaced autobiographies of religious conversion that were prevalent in earlier, more religious, periods (31). Hawkins notes similarities between these two types of autobiography, most notably that they both give "special prominence" to what she identifies as the "myth of rebirth"; that is, that the experiences of religious conversion and of illness afford "a process of transformation so profound as to constitute a kind of death of the 'old self' and rebirth to a new and very different self" (33).

Both MacDonald's and Swan's narratives describe their experiences of TB at around the time Hawkins pinpoints the emergence of pathography, and both exemplify the myth of rebirth that she delineates. MacDonald and Swan are removed from their everyday lives and sent to places from which they believe they might not return. Their emergence from the sanatorium, which signifies their unexpected survival of TB, is a kind of rebirth, but it is a rebirth that is haunted by the memory of those who did not survive. Although the publication history of these two texts is not the focus of my analysis, I want to touch on this history briefly because it reveals an important difference between the two narratives of tuberculosis: simply put, this history suggests which particular stories of illness can be told and who gets to tell them.[10] The belatedness of Swan's story in contrast to MacDonald's has everything to do with the public spaces available to tell particular stories, the dominant languages in which stories are told, and the impossibility of telling the stories of the literally countless Native American deaths from TB.

MacDonald's narrative was published not long after her sanatorium experience; initially parts of it were serialized in *Good Housekeeping* before being published in memoir form in 1948. When *The Plague and I* was published, MacDonald was already the well-known author of *The Egg and I* (1945), a humorous depiction of the trials and tribulations of life on a chicken farm. In contrast, Swan's account was not published until 1991, and she does not write her story but tells it to Mark St. Pierre, who, as the cover of the book explains, "has lived among the Lakota people since 1971, both as an educator and as an encourager of American Indian Art." Although the cover also makes clear that this is a Lakota woman's story "as told through" St. Pierre, he is listed in the Library of Congress classification as the book's author. As the person who framed the story, St. Pierre has had ultimate control over it. In his introduction, St. Pierre explains that he has arranged Swan's many stories "in a loose chronological order, creating the sense of a story line" (ix). *Madonna Swan: A Lakota Woman's Story*, then, is both Swan's "story" of tuberculosis and St. Pierre's "story line," a linear transformation of Swan's story.[11] St. Pierre also makes clear that this is not only Swan's individual story, but also the story of three generations of Lakota women and, even more generally, a twentieth-century American story. Through Swan's testimony the particular becomes universal, or as St. Pierre explains it, from the particular standpoint of a Native American woman who survived TB, Swan has a "keen ability to teach us—perhaps in a way we've never known—the twentieth-century experience" (ix). However, Swan's experience at Sioux Sanatorium is not at all a universal American sanatorium experience, but a distinctly Native American one.

There have been numerous histories of tuberculosis,[12] and some of these cite patient narratives describing the experience of tuberculosis.[13] TB also occupies a significant position in literature, particularly in the late nineteenth century, where, as Sontag (1978) has famously noted, it becomes a metaphor for expanded consciousness, artistic sensibility, and spiritual refinement. While such metaphorical qualities might create a distance between art and the experience of illness, a distance that Croce insists is necessary for good art, Sontag deplores the metaphorization and aestheticization of illness in art. This nineteenth-century understanding of tuberculosis, as somehow aesthetically and spiritually desirable, begins to change in the twentieth century as tuberculosis becomes a disease associated not with supposed upper-class refinement, but with urban poor and immigrant communities. Although the metaphors associated with TB shift from the nineteenth to the twentieth century, the sanatorium remains the space most closely associated with TB in the American public imaginary. As Katherine Ott notes in *Fevered Lives*, "[a]lthough only a small percentage of the tubercular ever spent time in a sanitarium, it became the representative twentieth-century site for tuberculosis" (1996, 136). Because most of those people able to write about their experience of tuberculosis were either middle or upper class, and, thus, could afford to spend time in a sanitorium, most fictional and autobiographical accounts of tuberculosis describe the experience of sanitorium life (see Rothman 1994, 215–40). MacDonald and Swan provide accounts of mid-twentieth-century sanatorium life that complicate our understanding of these representative sites for tuberculosis. The subjects of their narratives—working class and poor Native American women—as well as the depiction of what they undergo challenge dominant representations of TB and sanatorium life.

MacDonald's Tuberculosis Routine

In *The Plague and I*, the medical director at The Pines Clinic and Sanatorium diagnoses Betty MacDonald with tuberculosis based not only on what he sees inside her lungs but also on what he perceives of her disposition based on her outward appearance. MacDonald notes wryly:

> He said, "Taking the cure is going to be difficult for you. You have red hair—lots of energy, you're quick, active, impatient. All bad for tuberculosis. Discipline will be hard for you. The cure of tuberculosis is all discipline." I said that I would do anything. Anything at all to get well. He stood up and put his arm around me and said, "That is the spirit." Which was very kind of him considering the fact that he had just written on my card, "Prognosis—doubtful." (43)

The medical director diagnoses MacDonald's character as the opposite of that which defines patienthood; she is, the director discerns from her appearance alone, active and impatient not passive and patient, which makes her, he believes, unlikely to submit willingly to the discipline that a return to health requires.

The fictional Pines Clinic and Sanatorium is based on a clinic in Seattle and a Sanatorium called Firland in Richland Highlands, Washington, just north of Seattle, where MacDonald spent nine months in 1938.[14] Firland was not an elite institution that catered only to people who could afford to be there; indeed, many patients were from poor and working-class families who could not pay the fees. As MacDonald explains in *The Plague and I*, she is lucky to get a place at The Pines, and her account suggests that she may have been admitted because of a certain amount of health capital[15] that she accrues through her relationship with her brother-in-law, a pathologist. When she tells him, "[i]n a shaky voice and close to tears . . . about the diagnosis, the sanatorium and the $35 to $50 a week," he responds with the authority of a doctor and the assurance that comes with it: "The Pines, one of the finest sanatoriums in the world, is an endowed institution and free to anyone who needs care and cannot pay. There is a waiting list of over two hundred but mothers with small children are usually taken right in. I'll write a letter to the Medical Director" (34). When MacDonald meets the medical director for the first time, he is reading the letter that her brother-in-law has sent him about her particular case.

In her presentation of portraits of a diverse group of patients at The Pines, MacDonald reveals the racial politics of sanatorium life expressed through practices of spatialization. Although the majority of patients are white, there are several Japanese and African-American patients, including Kimi, a young Japanese woman who has an iconoclastic worldview, thanks, she believes, to her decidedly un-Japanese 5' 6" height, which makes her tower over other Japanese women and not fit in well, literally, to Japanese society. MacDonald learns that many of her fellow white patients will not room with the African-American or Japanese women. When the charge nurse tells MacDonald that she is going to put "Miss Sanbo [Kimi] in with you, if you don't mind," MacDonald is delighted and wonders why she should mind. The charge nurse explains that, "Some people would object to sharing a room with an Oriental," and MacDonald states simply, "I would prefer it" (130).

While the portraits of the patients and the nurses are drawn with precision and wit, the doctors are more elusive figures in MacDonald's account. She only encounters them periodically, and on these occasions there is no negotiation with her about her treatment. Rather, the doctors make decisions without consulting her, and the only sign from her doctors that she is improving in her time at The Pines is that her privileges gradually increase. She only learns that her sputum has been negative since shortly after her arrival at The Pines on the day that the medical director tells her she is well enough to leave The Pines for good. Despite the lack of dialogue with her doctors about her treatment, MacDonald's doctors are the heroes of this story. Their heroic status is first glimpsed in MacDonald's dedication. *The Plague and I* is "For Dr. Robert M. Stith, Dr. Clyde R. Jensen and Dr. Bernard P. Mullen without whose generous hearts and helping hands I would probably be just another name on a tombstone" (5). According to MacDonald, "[t]he staff at The Pines had but one motivating factor—to get the patients well" (60). She acknowledges that this motivating factor is "like a policeman's nightstick . . . twirled over our heads twenty-four hours a day" (60). But MacDonald believes that this is necessary because the patients themselves often don't want to get better. MacDonald calls this the paradox of a tuberculosis sanatorium: "It should be a place where the patients are striving to get well, aided by the doctors and nurses, but is actually a place where the patients are trying to kill themselves but are prevented, in many cases, by the doctors and nurses" (60). The constant struggle with the patient's own death wish inevitably makes the doctors and nurses less sympathetic to the patients and less willing to engage with them as individuals who happen to be ill. Although MacDonald is critical of the disciplinary regime practiced at The Pines and the impersonality favored by many of the nurses in dealing with the patients, she still agrees ultimately that this regime is for the patients' own good.

As *The Plague and I* demonstrates, MacDonald learns very quickly the moral meanings that are attached to disease in general and tuberculosis in particular. Indeed, she notices that the TB clinic where she is diagnosed shares a building with the police station, the city jail, the emergency hospital, and the venereal clinic; the link between criminality and disease is designed into the architecture of the building itself. The term "police," Foucault notes in his essay "The Politics of Health in the Eighteenth Century," was originally a generic word for the exercise of the functions of "order, enrichment and health" (1980, 170). The modern tuberculosis clinic and sanatorium might be conceived in Foucauldian terms as sites designed and "defined to correspond not only to the need to supervise, to break dangerous communications, but also to create a useful space" (1977a, 143–44). As Foucault describes it, then, the usefulness of the space is determined not necessarily in terms of the needs of the patients, but rather in terms of creating the most efficient organization for supervision.

The diagnosis of TB means that MacDonald will be considered in need of policing and supervision, not unlike those criminals housed in the city jail. She will be sequestered from the rest of society in order to break the potentially dangerous communication of a contagious disease. According to Foucault, "discipline proceeds from the distribution of individuals in space" (141); such distribution allows bodies to be more efficiently monitored and made "intelligible" under the gaze of authority. Thus, according to this schema, a tuberculosis sanatorium is an example of a "carceral" institution. "Unhealthy" tubercular bodies are sequestered from "healthy" bodies in the world outside the sanatorium and subjected to constant monitoring and normalizing judgments to determine their precise relationship to the norm of health. In her history and geography of tuberculosis, Ott explains that "[b]eing ill took place within a geographic space constituted by objects, tools, instruments, and people." She continues:

> In the case of consumption and tuberculosis, we can grasp the context of illness through the figurative and literal locations significant in their history: the sickbed and sickroom, the healing wilderness into which invalids retreated and its attempted reconstruction within the walls of the chest, the microscopic world of the bacillus, the consumer marketplace of invalid goods, the stereotyped black body, the Lung Block, and the sanitarium. These are distinctive sites of tuberculosis—spaces inhabited by living beings and shaped by material objects. Patients, practitioners, and the community came together to build the optimum environment for the illness and thus to define it and fix its identity within these spaces. (1996, 5)

Tuberculosis as experience and event comes into being through the various spaces, objects, and people associated with its diagnosis and treatments. We will see when we take up Swan's account of TB in the Native American community that there are multiple sites of tuberculosis, and that not all sanatoriums are sites where patients are prevented from, in MacDonald's terms, killing themselves by the heroic will and expert knowledge of doctors and nurses. The "optimum environment for the illness," as Ott understands quite well, changes over time and across communities.

In his genealogical work, Foucault identifies modern forms of power as not (or not only) repressive, but rather as productive, forming a "positive economy" of power/knowledge. Power is not exerted unilaterally from a single position above (for example, the position held by the "sovereign" or the "state") downwards in an exclusively hierarchical manner; power is instead *dispersed* multilaterally, and its exercise, though often invisible, permeates and, indeed, constitutes individual bodies within the social body. Discipline, according to Foucault, is the means by which modern power operates. In *Discipline and Punish*, he asserts that discipline "is a type of power, a modality for its exercise, comprising a whole set of instruments, techniques, procedures, levels of applications, targets; it is a 'physics' or an 'anatomy' of power, a technology" (1977a, 215). And in a number of his works, Foucault identifies health care as a primary technology in which the individual is subjected to discipline (1973, 1977a, 1980).[16] Health care, in other words, is a form of discipline in which the cared-for body is a hyper-managed, ever analyzable, and, thus, "docile" and "intelligible" body.

Where health is concerned, documentation forms the basis for differentiation, classification, and segregation. As Foucault notes in *Discipline and Punish*, "The examination that places individuals in a field of surveillance also situates them in a network of writing; it engages them in a whole mass of documents that capture and fix them" (189). At The Pines, for example, the temperatures of the "inmates" are taken at regular intervals, and the results are diligently recorded on charts, the slightest rise in temperature indicating a movement away from health, and the slightest decrease indicating a movement toward health. When MacDonald arrives at The Pines for her indefinite incarceration, a nurse goes over the rules—the "tuberculosis routine" (1948, 94)—she must follow in order to be allowed to stay. For these are rules one must willingly submit to in the hopes of getting better:

> Patients must not read. Patients must not write. Patients must not talk. Patients must not laugh. Patients must not sing. Patients must lie still. Patients must not reach. Patients must relax. Patients must . . . (53, ellipses in original)

Patienthood, then, is not reading, writing, talking, laughing, singing, or, even, reaching; it is lying still, relaxing, and, most of all, obeying the rules. The norms of health are arrived at through surveillance by health care professionals, and by the patients themselves, through the interiorization of the medical gaze, the willing commitment to the "tuberculosis routine," and the implementation of the most meticulous self-monitoring. The ability to monitor oneself effectively indicates one's desire for health, and one's desire for health indicates one's fitness for citizenship in a normalizing society. One of the lessons of the tuberculosis routine, MacDonald explains, is that, in order to continue to receive the much-sought-after care that The Pines provides, patients must always be on their best behavior and "grateful to the nurses, the doctors, and to the sanatorium" (122). As MacDonald notes with irony, the final point of this lesson is a moral one that requires disciplining the body by disciplining the mind: "IF YOU THINK RIGHT, YOU WILL ACT RIGHT" (122; emphasis in original).

In his discussion of the constitution of docile bodies in *Discipline and Punish*, Foucault gives, as an example, methods for the "correct training" of the soldier. For the soldier, detailed exercises are imposed such that "[a]ll the activity of the disciplined individual must be punctuated and sustained by injunctions whose efficacity rests on brevity and clarity, the order does not need to be explained or formulated; it must trigger off the required behavior and that is enough" (1977a, 166). Foucault detects such forms of *dressage* not only in the particular situation of soldiers in military training, but also in general as the means by which modern power operates. At The Pines, however, the exercises the patients are subjected to are, paradoxically, *forms of rest*. In fact, according to the official code, "everything that is not rest is exercise" (1948, 72). But, like Foucault's forms of *dressage*, this constant rest requires an infinite amount of attention to the body and the self: again, no talking, no laughing, no coughing, no singing, no reaching. What is required, it seems, is virtually no breathing at all; in fact, the closer to dead one is, paradoxically, the better one is, or, at least, the better one is for monitoring.

The "tuberculosis routine" that MacDonald and her fellow patients *endure* is not only enforced while they are at The Pines, but is extolled as what Foucault would call an "art of existence" or "technology of the self," which he defines as "forms of *elaboration*, of *ethical work (travail éthique)* that one performs on oneself, not only in order to bring one's conduct into compliance with a given rule, but to attempt to transform oneself into the ethical subject of one's behavior" (1985, 27). In *The Plague and I*, MacDonald tells us that the most important lesson taught at The Pines, according to the rule book, is: "The cure of tuberculosis is not medicine but a new regime of living, not only during the sanatorium period, but for years and years, maybe for a lifetime afterwards" (128). It is this *new regime of living* promoted in the name of health that I am concerned with here, in terms of both its hegemonic exercise and the possible disruption of its exercise. What characterizes this hegemonic regime of living, and how might this hegemonic regime be transformed and/or transgressed?

At The Pines, MacDonald learns quickly, the nurses are in charge of enforcing discipline, and "impersonality" between patients and nurses is mandated; such "impersonality" maintains and bolsters the border between health and illness, the normal and the pathological (45). In *The Social Transformation of American Medicine* (1982), Paul Starr argues against a critique of public health that posits public health clinics and the nurses and others who work at them as "agents of social control." Although Starr admits that public health clinics were involved in transmitting "middle-class American standards of value" as much as "scientific information," nonetheless, he asserts that "there can be little question that the new hygienic practices had value in preventing disease and preserving health" (192). While I think Starr is correct in pointing out the important disease-preventing work of public health clinics, it might also be said that a Foucauldian analysis of health would not describe tuberculosis clinics and sanatoriums like The Pines and the nurses and others who work at them as "agents of social control," but rather would understand them as caught up in the same nexus of power/knowledge as their patients. The nurses at The Pines, then, are neither agents of social control nor agents of reform; rather, the reforms they articulate and implement are imbricated in the disciplinary apparatus of health.[17]

Although power, in Foucault's formulation, is not hierarchical per se, nonetheless, its normalizing function does create hierarchical binary relations: between doctor and patient, doctor and nurse, male and female, rich and poor, white and nonwhite, citizen and immigrant. Despite these normalized hierarchical binary relations, the new regime of living is recommended not only for the patients at The Pines, but for everyone. The nurses at The Pines, therefore, are subject to the same discipline as their charges:

> Molly told us some of the trials of being a nurse at The Pines. She said that the discipline was not limited to the patients as the nurses were not allowed to smoke on the premises, had to be in every night by ten-thirty, were required to attend school three nights a week and were under twenty-four-hour surveillance to be sure that they obeyed these rules and many others, including no indulgence in SEX, thoughts of SEX, actions which might eventually lead up to SEX, discussions of SEX, or literature concerned with SEX. She said that with the exception of the charge nurses, the nurses weren't allowed to speak to the doctors, which made it rather difficult as she and one of the staff doctors were engaged. (125; emphasis in original)

Note the explicit link between health and sexuality in this paragraph. In his overturning of the "repressive hypothesis" in volume 1 of *The History of Sexuality*, Foucault contends that, as is demonstrated here, sex isn't actually repressed in the late-nineteenth and early-twentieth-century West, but obsessively returned to in discourse. This quote from MacDonald is interesting, therefore, with regards to modern technologies of health and sex, both of which Foucault would call "techniques for maximizing life" (1978, 123). Both health and sex are "transfer point[s] for relations of power," and, in MacDonald's description, the two converge in the figure of the nurse, who might be conceived of as a liminal (and, perhaps, problematic) figure between doctors and patients. MacDonald's ironic portrait reveals as well what Foucault has called *scientia sexualis*, or procedures for telling the truth about sex, which are in operation even when sex itself is being inveighed against. Although everyone is subjected to this new regime of living, as MacDonald's discussion of The Pines nurses reveals, not everyone experiences it in precisely the same way. Those who are deemed outside the norm—by virtue of gender, race, class, disability, and/or illness—are subjected to greater discipline.[18]

The Plague and I presents an image of the patient as constructed within the institution of modern medicine and through the practices of the anatomo-clinical method and the disciplining of individual bodies in space. MacDonald herself knows the ideal image of the patient, and although she attempts to conform to it, at times she fails to do so. Her acknowledgement of the inevitability of failure in conforming to the tuberculosis routine is a form of subjugated knowledge that reveals the intrinsic absurdity of the patient's position, and might constitute a proto-ethics of failure.[19] Through humor, MacDonald challenges the patient's position as constructed through the anatomo-clinical method. Yet, despite MacDonald's sometimes irreverent rhetorical stance, nonetheless, she readily submits to medicine's techniques of discipline and surveillance in order to prove that she has both a healthy body and a healthy disposition. Moreover, although to some extent she challenges the relationship between tuberculosis and criminality, simply by making that implicit relationship explicit to her readers, she also accepts—indeed, she understands that she must accept—the conventional moral meanings ascribed to health and illness. While she recalls certain noncompliant patients with humor, she also makes it clear to her readers that their behavior is preventing them from getting well. In *The Plague and I*, MacDonald attempts to de-link tuberculosis from criminality in her individual case, rather than more broadly. Hers is a personalized defense in the memoir form rather than a defense that seeks to undermine the binaries—health and illness, normal and criminal—that structure how we think and know the experience of tuberculosis. MacDonald's work, then, is both hegemonic and transgressive because she both submits to the tuberculosis routine and subverts it. On one hand, the memoir brings the tuberculosis routine to the reading public at large and promotes it as a way of life, an art of existence. On the other hand, MacDonald's irreverent treatment of the routine as it was promulgated and lived at The Pines is a form of resistance to it. With humor, MacDonald undermines the sanctity of the routine even while she extols its virtues and attempts to uphold its rules. Yet, despite the initial concern with her supposedly unpatient-like disposition, which seemed to disadvantage her at the outset, the moral of MacDonald's story is that she has followed the routine so well that she gets out of The Pines alive, and apparently in record time.

Madonna Swan's Death Diary

Unlike MacDonald's often irreverent portrait of sanatorium life, the story of tuberculosis that Madonna Swan tells to St. Pierre, who then crafts it into a "story line," is introduced as a heroic overcoming narrative of an individual and her community. For St. Pierre, "this collection serves as a tribute to all those Native Americans who survived tuberculosis, as Madonna did, and represents a closing and healing of that terrible wound in Lakota and human history, a period darkened by the sheer magnitude of an epidemic whose impact on tribal societies has been largely overlooked in Native American literature" (1991, x). While Swan's story of tuberculosis is situated within the larger and longer story of her family and her tribe, the centerpiece of the book is her chilling account of her experience as one of the many Lakota children and adults who contracted TB and were removed from their families, sometimes forcibly, and put in sanatoriums exclusively for Native Americans. In a study on the relationship between public health nurses and clients on Indian reservations in the 1930s, Abel and Reifel note that "most state and county sanatoria refused to admit American Indians," and so Native American tuberculosis patients had to be placed in institutions established first by the Office of Indian Affairs and later run by the United States Public Health Service (1999, 497; see also Raup 1959). These institutions, according to Abel and Reifel, "were seriously deficient" (498), and Sioux Sanatorium, where Swan was kept from 1944 until she "escaped" in 1950, was no exception.

After a prairie fire that she and her fellow students at the Indian boarding school are called on to help fight, Swan develops a persistent cough. She loses weight, grows gradually weaker, has chest pains, and becomes susceptible to colds. Eventually she becomes so weak that she cannot get out of bed. A Sister at her school asks her why she is "so down," and she tells the nurse, "I think I'm going to die like Rita, Rosie, and those others" (61), all of whom have died of tuberculosis, a "word that [Swan] would hear so often and come to hate with [her] whole soul" (55). A doctor examines Swan, taking a sputum test and lung x-ray. He tells her, "All of your tests are positive and your lungs are all clouded over. That could be damage from the prairie fire, but you do have tuberculosis" (61). This is the "awful dreaded word" Swan does not want to hear. Among Native Americans at this time, a diagnosis of TB was frequently equivalent to a death sentence.

By the time she gets her own diagnosis, Swan has already witnessed several TB deaths, and she associates TB not only with an actual and horrible death, but also with a social death, because it incurs shame and stigma on individuals and entire families within the Native American community and beyond. Like MacDonald, she too understands the stigma attached to those who are sent to a sanatorium. However, the image of the sanatorium that Swan presents is quite different from the image of the sanatorium in MacDonald's memoir. Swan is less concerned with the association between TB and criminality than with the fact that the Sioux Sanatorium (Sioux San) run by the United States Public Health Service in Rapid City, South Dakota, is a place where Native Americans go not to be treated, but to die.

In the last chapter of volume 1 of *The History of Sexuality*, as well as in the recently published lectures given at the Collège de France in 1975–76, Foucault returns to the anatomo-clinical method and disciplinary forms of power in order to supplement this analysis of "power over the body in an individualizing mode" with an analysis of a second mode, which he calls a "massifying" mode that is "directed not at man-as-body but man-as-species" (2003, 243). The first mode, as I described earlier in this chapter, is what Foucault, in March 1976, calls an "anatomo-politics of the human body," and the second mode, which does not replace but complements the first, is a "'biopolitics' of the human race" (243). The approach to tuberculosis prevention and treatment in the early to mid-twentieth-century United States, just prior to the discovery of antibiotics that have been successful in treating TB, in certain communities at least, shows both of these modes at work.

In his lectures, Foucault describes the transformation in the nineteenth century from a right of sovereignty to "take life or let live" to a new state right to "make live and let die" (241). In the case of tuberculosis prevention and treatment in the twentieth century, we see this new state right demonstrated: *The Plague and I* presents the right and practice of making live, while Swan's illness narrative presents the right and practice of letting die. MacDonald's memoir records the making live mode of population control, as well as the disciplinary power over individual bodies; Swan's illness narrative records the letting die mode of population control, in that the Native American sanatoriums are less concerned with helping people to live, and more concerned with letting them die—away from the larger population.[20] Here we see exactly the relation between biopower and racism that Foucault describes in his lectures. The power to let die, which is deployed against particular populations, is rationalized through racial hierarchies. For Foucault, "The fact that the other dies does not mean simply that I live in the sense that his death guarantees my safety; the death of the other, the death of the bad race, of the inferior race (or the degenerate, or the abnormal) is something that will make life in general healthier: healthier and purer" (255). Foucault locates "[s]exuality... at the point where body and population meet" (252), but, of course, it is not only sexuality that must be disciplined and regularized in order to make live and let die, but also disease, which is, like sexuality, understood in terms of both individual bodies and populations. According to Foucault, "Medicine is a power-knowledge that can be applied to both the body and the population, both the organism and biological processes, and it will therefore have both disciplinary effects and regulatory effects" (252).

In MacDonald's description of the tuberculosis routine, we see disciplinary effects on individual bodies, and, because this routine is meant to be taken up by everyone in society, we can also see that it is meant to have a regulatory effect on an entire population as well. Although Swan was also subjected to forms of discipline at Sioux San, I want to focus here on the methods by which the Native American population, or at least Swan's Lakota Sioux tribe, was subjected to forms of regularization, which amounted to letting those with TB die. One might think of the Native American population in the United States as *regularized out of existence* through both assimilation—and the Catholic boarding school where Swan contracted TB is one such space for this form of regularization—and diseases, like TB, which were allowed to run their supposedly "natural" course through entire Native American populations.

We might think of tuberculosis in the Native American community as endemic not epidemic, if we take account of the distinction between the two that Foucault makes in his 1976 lectures. Endemics were "the form, nature, extension, duration, and intensity of illnesses prevalent in a population," according to Foucault. "These were illnesses," he continues,

> that were difficult to eradicate and that were not regarded as epidemics that caused frequent deaths, but as permanent factors which—and this is how they were dealt with—sapped the population's strength, shortened the working week, wasted energy, and cost money, both because they led to a fall in production and because treating them was expensive. In a word, illness as phenomena affecting a population. Death was no longer something that suddenly swooped down on life—as in an epidemic. Death was now something permanent, something that slips into life, perpetually gnaws at it, diminishes it and weakens it." (2003, 243–44)

Although he was speaking more generally, in his delineation of the difference between endemic and epidemic diseases, Foucault could have been describing tuberculosis in the Native American community in the first half of the twentieth century.[21] In utilizing Foucault's understanding of endemic disease for analyzing TB among Native Americans, it is important to keep in mind that TB became endemic in this population only after contact with whites. As Georgina Feldberg notes in *Disease and Class*, studies in both Canada and the United States in the early twentieth century showed a steady "tuberculinization" of Native Americans because of increased contact with whites (1995, 162).[22]

While the possibility of death hangs over MacDonald's work, it is not pervasive and permanent in the manner Foucault describes. She admits before she goes to the sanatorium that in her mind's eye "[s]antoriums were places in the Swiss Alps where people went to die. Not only that but everyone [she'd] ever heard of who had had tuberculosis had died" (33). And yet, once she arrives at the sanatorium, death is not as prominent a feature of life for her at The Pines as it is for Madonna Swan at Sioux San. This is partly because those patients at The Pines who are very ill are isolated from the rest of the patients, and partly because the tuberculosis routine requires a positive attitude, which doesn't include dwelling on death. As the charge nurse tells MacDonald, "We do not allow patients of The Pines to think about death, or other unpleasant things. You must have pleasant cheerful thoughts" (161). At Sioux San, on the other hand, all patients—those who are very ill and those who are less ill—are kept together. Swan speculates, probably correctly, that, "those of us who were highly positive kept reinfecting those that had negative sputum cultures" (73). Perhaps revealing the possible benefits of the disciplinary mode that would have separated very sick bodies from less sick ones, Swan notes, "They should have divided or separated us so we didn't reinfect the healthier ones, but they didn't. I felt bad about that" (73). At Sioux San, the Native American patient population was massed together, seemingly not to prevent death but to allow it.

Both MacDonald and Swan describe the sanatoriums they are sent to as prisons, but MacDonald's experience as an inmate at The Pines is merely figurative, nothing is keeping her there but her own desire for a return to health, and, in fact, the main threat hanging over the patients at The Pines is that those who do not abide by the tuberculosis routine will be sent home. Swan's experience as an inmate is much more literal.[23] At the Sioux San, at least in its early years, the patients were "dressed all alike, like the inmates at the penitentiary in Sioux Falls, all dressed in stripes." Swan explains further, "I guess that was intended to keep us from escaping. If we could escape, we could not get far. Of course, our regular clothes were locked away" (74). Although the patients could stand on balconies to get some fresh air, they were not allowed to set foot outside the buildings.[24] Swan understands this in terms of the stigma attached to diseased bodies:

> Living in the san would make you feel like an outcast with some filthy disease like leprosy. We couldn't go outside. We were allowed only to stand out on the little balconies and look out across Rapid City, watching people go about their daily lives, enjoying life. From 1944, through 1945 and 1946, until 1947, we were not allowed to go outdoors, not stand on the ground, Maka Ina, Mother Earth, even once. Just going to the window or a balcony was all we could do, but even that was a relief—at least it was fresh air. (80)

We might also understand this literal incarceration as a form of regularization in which an always already-diseased Native American population is kept separate from a supposedly nondiseased white population. As I've already noted, the irony of this healthy white/diseased Native American binary is that, in the case of TB, it was Native Americans who needed protection from whites.[25]

In her first week at the San, Swan meets an old friend, Margaret Halfred, who is very ill. Rather than feeling comforted by seeing a familiar face in an unfamiliar place, Swan is discomfited because she sees her future in Margaret: "just seeing what she had become made me fearful" (71). In fact, Margaret is the first person to die after Swan's arrival at Sioux San. Swan learns quickly why Margaret has been moved to a private room; those who are dying are moved out of sight to a part of the sanatorium that the patients call "death row." Being on death row does not mean, however, that one is formally isolated from the other patients. One of Swan's tasks at the sanatorium is to bathe women who are too sick to bathe themselves, and to wash the hair of women on death row. Swan explains, "We always helped out with the sick ones, I guess, because they were short of staff" (75). By caring for those who are sicker than she is, Swan risks reinfecting herself, and at the same time must face again and again a death that she senses will be her own.[26] This is not just pessimism on her part, but reality for the patients at Sioux San:

> There were no patients going out on leave or dismissed in 1944 or 1945. Not until 1946 did the first few patients leave alive. Death was the only way anybody left before that, and there were many. I had a diary, and in it I wrote down the little things that happened each day, things girls outside the san would never have bothered to write down. I wrote down whatever happened, no matter how ordinary, things like the kind of day it was, how I felt, and if an outsider came to visit. I wrote down why and when they came, and I wrote down the names of those that had died. (74)

Swan's diary, unlike MacDonald's memoir, becomes a diary of death. She and another patient and diary keeper, Bernice Long, review their diaries together in 1950 and count five hundred deaths. Swan and Long note a pattern: "It seemed that when one would die, two more would die soon after. That was something both of us had noted in our diaries. When one would die, we would wonder who the other two would be. We wrote that at varied times, always wondering if the next one would be us or one of our friends" (74).

After six years at Sioux San, the doctor there, Dr. Sedlechek, still insists Swan cannot leave. When she asks if she will soon be able to leave, he tells her, "No, your tests are very bad. Your sputums are bad and your x-rays are bad! I can't let you go home because your germs are very bad" (101). In 1950, Swan's younger brother Orby also becomes sick with TB and joins her briefly at Sioux San. But from the beginning he has no intention of staying. He tells his sister, "I'm not going to be like you. They're not going to keep me here and not let me go home. I know I'm going to die. I want to die at home" (101). Orby simply tells his father to come pick him up, and he walks out of Sioux San. When he dies at home shortly thereafter, Swan is unable to leave to attend his funeral. Haunted by her brother's words and actions, Swan decides to escape, and just the idea of escape gives her life new meaning. She arranges to have someone meet her at the sanatorium with a car, and even when she is being driven away from the Sioux San, she fears that "the Rapid City police will arrest [her] like a common criminal" (104). There are reprisals for her escape; a red tag is put on her home marking it as a "TB house" (104). But her father decides she will not go back to Sioux San, and he begins to use connections to try to get her into the white sanatorium known as Sanator. When her father is told that Madonna cannot go to Sanator because Sanator is for whites and Sioux San is for Indians, he protests that Indians fought bravely for the United States in World War II, including two of his sons, one of whom died in battle.

Swan is eventually admitted into Sanator, and the difference is dramatic: "The grounds were pretty, with trees and flowers and all. The patients were walking the grounds with their own clothes on" (108). But, even more than the fact that it is less prison-like than Sioux San, Sanator has doctors who apparently want to make Swan better, not just let her die. Swan is given a new experimental surgical treatment, in which ribs and upper and lower lobes of one lung are removed (113). She agrees to this experimental surgery in the hopes of helping other people with TB in the future as much as to help herself in the present. After the surgery, Swan's recovery is arduous: she is bedridden for six months because she can't support her spine and head, and is numb from her fingertips to her shoulder (117). Her doctor is grateful for what he learns from her surgery, and tells her, "We are already making many changes in our technique because of you and those other brave ones who went first" (121). Miraculously, Swan does finally begin to recover, but apparently not because of the experimental surgery. Just before the close of the midsection of her oral history, which describes the ten years she suffered from TB, she mentions that "[t]hey had put me on antituberculin medicine, and for the first time since 1944 I knew I would be cured" (121).

In the end, Swan herself, with the help of St. Pierre, presents a narrative not only about overcoming TB but also about overcoming racism. It is Sanator and its white doctors that save her, and the grateful former patient continues to work as a receptionist at Sanator even after she has recovered from TB. Swan's story is a story of survival: if she had not gone to Sanator when she did, if she had not been given antibiotics when she was, she very well might have died, and her oral history would never have been published for others to read. This is the dominant narrative that she tells, but this narrative paradoxically both covers over and reveals the racism that led to the countless Native American deaths from TB. Within the biopower mode as Foucault describes it, racism becomes the justification for "the need to kill people, to kill populations, and to kill civilizations" (2003, 257). Foucault maintains, "When I say 'killing,' I obviously do not mean simply murder as such, but also every form of indirect murder: the fact of exposing someone to death, increasing the risk of death for some people, or, quite simply, political death, expulsion, rejection, and so on" (256). Madonna Swan's story may be about overcoming, but it is haunted by another story of racism that "justifies the death-function in the economy of biopower" (258). What we hear in Madonna Swan's testimony is an echo of the diary of death she kept at Sioux San, which is an echo of all the Native Americans who were killed there.

2.

Politicizing Patienthood

Ideas, Experience, and Affect

In the final two volumes of his three-volume *History of Sexuality*, Foucault turns from his earlier analyses of discursive practices (his archaeological work) and the "manifestations of power" (his genealogical work) to an analysis of the constitution of the subject, or what he calls the "arts of existence" or "techniques of the self" (1985, 6–11). In *The Use of Pleasure* and *The Care of the Self*, Foucault looks at the particular arts of existence in Classical Greece and Imperial Rome, and the transformation of these arts from one period to the next. If in the last chapter I was interested in Foucault's theory of power, his genealogy of the subject interests me most here; he approaches the history of sexuality not, or not only, through an analysis of institutional codes or juridical authority but through forms of subjectification, or practices of the self, that included in Classical Greece, for example, an ethics based on moderation, or, as Foucault explains, the "domination of oneself by oneself" (1985, 65). Nikolas Rose describes Foucault's (as well as his own) methodology as a "genealogy of subjectification" whose "domain of investigation is that of practices and techniques, of thought as it seeks to make itself technical" (1996, 128). For Rose, a genealogy of subjectification looks at the "kinds of attention that humans have directed towards themselves and others in different places, spaces and times" (129), in order to achieve certain "ethical" objectives associated with such categories as "manliness and femininity, honour, modesty, propriety, civility, discipline, distinction, efficiency, harmony, fulfillment, virtue, pleasure" (130). To this list that Rose provides, I would add the "ethical" objective of health.

I see memoir in general as a form of subjectification, a practice in which we consider our relationship to ourselves, and illness narratives in particular as describing the cultivation of a self in relation to itself and to others, healthy, ill, and in-between. Foucault asks that we think "in terms of a crisis of the subject, or rather a crisis of subjectivation[1]—that is, in terms of a difficulty in the manner in which the individual would form himself[2] as the ethical subject of his actions and efforts to find in devotion to self that which could enable him to submit to rules and give purpose to his existence" (1986a, 95). As I showed in my discussion of the two TB narratives, illness enacts a transformation, a crisis of the subject, a crisis of subjectification itself. The transformation enacted depends, of course, on the particularities of the illness—whether, for example, it is an acute illness or a chronic illness; whether it is an infectious or noninfectious disease; whether it affects the physical body or the mind or both; to what degree and in what manner it stigmatizes the individual who is ill as well as on the particular sociocultural milieu in which the ill person is situated. In this chapter, I consider several different forms of subjectification that illness enacts—what I call *arts of being ill or doing illness*—in the United States at a particular moment in time, the last two decades of the twentieth century. Influenced by Foucault's description of his own project, my work seeks "to ascertain the set of transformations in the regime of discourses necessary and sufficient for people to use these words rather than those, a particular type of discourse rather than some other type, for people to be able to look at things from such and such an angle and not some other one" (1980, 211).

I want to explore the ways that a particular illness in a particular place and time—in this case, breast cancer in the United States at the end of the twentieth century—becomes an illness event. The emergence of breast cancer as an illness event must be understood in connection with two other events related to illness: feminist health activism of the 1970s and AIDS activism of the 1980s. Unlike the narratives of TB, where illness remained for the most part a personal experience framed by medical knowledges (though with significant sociocultural effects) all four of the illness narratives I discuss in this chapter are influenced by these two events that constitute illness as political, and all of them describe practices of subjectification that give rise to the "politicization of patienthood." I take this term from feminist philosopher and "theorist of plagues"[3] Linda Singer, who called for a politicization of patienthood by persons with cancer in the manner of AIDS activists.[4] For

Singer, the politicization of patienthood resists the "isolation imposed by illness" and insists on "maintaining some sense of community and constituency" among persons with cancer (1993, 105). According to Singer,

> AIDS rhetoric and politics acknowledge, in a way that usual medical and public health discourse does not, the anger and frustration of contracting a disease for which there is no existing cure, and where medical knowledge and practice are inconclusive and indecisive. Throughout the rhetoric of AIDS politics, there is an important rejection of religious logic and a religious imagery, the hope for salvation or apocalyptic reprieve. There is a rejection as well of the posture of placing faith in doctors, searching for another Lourdes, "being a good patient" or pious recipient who accepts whatever is offered gratefully and without question. (106)

In this chapter I look at several forms of politicization that emerge in the fifteen years between 1978 and 1993, and that link the women's health movement of the 1970s to the AIDS movement of the 1980s and 1990s and the breast cancer movement of the 1990s. I believe that the politicization of patienthood brings into being various techniques for doing illness in new ways, and that along with these new forms of doing illness come new forms of writing illness.

I begin with an emergent politicization that can be discerned in Susan Sontag's classic discussion of metaphor and illness, and her assertion that all metaphorical thinking must be banished from our response to illness, In 1978, Sontag published a series of essays in the *New York Review of Books* that would later become the bestseller, *Illness as Metaphor*.[5] Although at first glance Sontag's work does not appear to be a personal response to her own experience of illness, I argue that it might be read as, paradoxically, a depersonalized personal narrative of illness. In fact, Sontag depersonalizes, and also de-heroicizes, her response to illness in order to, in her view, offer a strategy to others that she believes is most *effective* in the face of illness. I next consider Audre Lorde's *Cancer Journals* (1980), published by a small feminist press just two years after *Illness as Metaphor*.[6] I take 1980 to mark the emergence of a new figure—the politicized patient—and a new genre—the patient's counternarrative to medical discourse as exemplified by doctors' charts and case histories. Since 1980 we have seen the proliferation of both activism around health issues as well as personal writings about the experience of illness from the patient's perspective, and I consider Lorde as an exemplary writer-activist who transformed the silence surrounding breast cancer through language and action. I then analyze Eve Kosofsky Sedgwick's autobiographical essay "White Glasses," first presented at a conference at the CUNY Center for Lesbian and Gay Studies in 1991 and later published in the collection *Tendencies* in 1993. Like Sontag and Lorde, Sedgwick also wants to challenge the hegemonic sick role and its passively heroic mode of being ill, and she does so by making use of her theoretical and political concept of "queer performativity" in relation to her personal experience of breast cancer.

All three writers I discuss here challenge the structures and structuring of illness from the patient's side of the doctor-patient binary; they all present effective histories that are attentive to the rhetorics and practices of politics. They diverge from each other in their relationship to the affective and its place in their illness narratives. For Sontag, at least initially, the affective has no place in accounts of illness. She believes that by purifying the language we use to speak and write about illness of both metaphor and of the affective that often gets expressed through metaphor, we will transform the experience of illness itself in necessary ways. Unlike Sontag, Lorde and Sedgwick are not interested in purifying the language with which we speak and write about illness; they believe that the personal and affective has a central place in their politics and writing, and in how they do illness. While Sontag's challenge to conventional treatments of illness is formulated in terms of an intellectual idea that has rhetorical and proto-political effects, Lorde's is formulated in terms of a personal experience that has poetic and political effects, and Sedgwick's is formulated in terms of an affective experience that has queer performative and political effects. Or, put somewhat schematically, Sontag's politicized patient does things with ideas, Lorde's does things with experience, and Sedgwick's does things with affect.

Although the politicization of patienthood is at the center of this chapter, and indeed at the center of *Treatments*, I argue that in the 1990s the emergent politicization that can be discerned in the illness narratives of Sontag, Lorde, and Sedgwick gets covered over by a neoliberal mode of being ill and doing illness that emphasizes the discourses and practices of personal responsibility in matters of health. Although this mode of being ill has some kinship to the tuberculosis routine and its forms of discipline that I described in the last chapter, it also operates somewhat differently coming as it does after the period in which the discourse of politicization of patienthood emerges.[7] In order to delineate the discourses and practices of responsibility that coexist—or indeed cover over—the discourses and practices of politicization, I turn in the final section of this chapter to Barbara Ehrenreich's widely circulated essay, "Welcome to Cancerland" (2001), which investigates how and why this covering over of a radical political approach to health happened in the last decade of the twentieth century, and critiques what she calls "breast-cancer culture" and its intimate relationship with, rather than challenge to, the "Cancer Industrial Complex."

Doing Things with Ideas

Ironically, it was a book "against interpretation" that would play an important role in a shift away from the concern for sickness as role and disease as diagnosis to a conception of the illness experience as a whole complex of forces and relations. In *Illness as Metaphor,* Sontag critiques the social and moral meanings that are attached to certain illnesses; she problematizes the metaphorization of illness and wants to "de-mythologize" disease. To some extent, Sontag's desire to render illness as devoid of moral meanings is not unlike the anatomo-clinical approach, which attempts to render illness as a pathological fact, an object of analysis for medicine's "speaking eye." Her appeal is patient-centered, but in such a way that diagnosis is of disease as an entity separate from a patient's personality or disposition, which is the opposite of what we saw in MacDonald's case, where the clinic doctor initially diagnosed her as someone unlikely to recover from TB because of what he perceived as her undisciplined disposition. In her work, Sontag attempts to delink illness from normalizing judgments. Tellingly, though perhaps not surprisingly considering its date of publication, Sontag's work mentions few patients' autobiographical narratives, focusing rather on fictional and medical representations.

By comparing the representations of tuberculosis in the nineteenth century and the representations of cancer in the twentieth century, Sontag suggests that diseases whose etiologies are unknown are most likely to be metaphorized in both medical and popular understandings. In the nineteenth century, according to Sontag, TB was a disease that was romanticized, and as such represented not so much as a debilitating disease, which it clearly was, but as an opportunity for "spiritual refinement" and "expanded consciousness."[8] Unlike TB, cancer has never been romanticized, Sontag asserts, nor has it been aestheticized. In the twentieth-century representation of cancer, the disease becomes not a reflection of the sufferer's spiritual refinement, but, instead, a reflection of the sufferer's allegedly repressed character. Cancer does not expand consciousness, but obliterates it. Sontag argues that these metaphors—both the good nineteenth-century TB metaphors and the bad twentieth-century cancer metaphors—are damaging for those persons who are suffering from the actual diseases, which in and of themselves, she insists, do not have moral meanings. Thus, Sontag is impatient with the need to make illness meaningful, even, or especially, by attributing to the experience of illness the impetus to change one's life, to make it meaningful in ways it wasn't before.

Interestingly, considering Sontag's stated de-metaphorizing and de-mythologizing agenda, many commentators on Sontag's work have seemingly misread it, often citing only the metaphorical image with which she opens *Illness as Metaphor*:

> Illness is the night-side of life, a more onerous citizenship. Everyone who is born holds dual citizenship, in the kingdom of the well and in the kingdom of the sick. Although we all prefer to use only the good passport, sooner or later each of us is obliged, at least for a spell, to identify ourselves as citizens of that other place. (1978, 3)

Such emigration/citizenship/national character metaphors have become a recurrent trope in personal narratives of illness; illness is often referred to as another country to which one is temporarily or permanently exiled.[9] Which is more useful, then, Sontag's metaphor or her argument against metaphor? I want to emphasize that both—the de-metaphorizing idea and metaphorical language—might be useful for the person who is ill, and both have been crucial in transforming the way illness is spoken.

One problem with Sontag's argument is that in asserting that "the healthiest way of being ill" (3) is experiencing illness purified of metaphorical thinking, she retains the health/illness binary characteristic of modern medicine. She seems to be saying, if only implicitly, something that Croce, MacDonald, and Swan might also be said to have articulated, that there are good and bad ways not only to be ill, but also (and this is especially important to Sontag, as it is to Croce) to write about or represent being ill. In her earlier essay "Against Interpretation," Sontag asserts that she wants, and believes it is possible to have, "pure, untranslatable, sensuous immediacy" in art (1966, 9). In the same essay, she explains further that she is interested in the "sensuous surface of art" rather than in "mucking about in it" (13). We might ask, however, what is the difference between "the sensuous immediacy" of an experience and "mucking about in it"? I contend that many illness narratives, in order to describe the sensuous immediacy of illness, must in fact muck about in it. The "it" that they muck about in is not only the experience of a body that can no longer be taken for granted, but also the affective responses to that body as it becomes undisciplined.

Sontag's desire for an art that is "unified and clean" is in contradiction, it seems, to her assertion that "[w]hat is important now is to recover our senses. We must learn to *see* more, to *hear* more, to *feel* more" (14). In the ways one responds to art as well as in the ways one responds to the experience of illness (and, of course, the attempt to give this experience of illness form through art), Sontag is positing the possibility of some prediscursive experience.[10] But, interestingly, this prediscursive experience, in Sontag's conception, is not unlike what might be conceived of as an objective or scientific experience: "detached, restful, contemplative, emotionally free, beyond indignation and approval" (27). Sontag is intent on distinguishing the sensual from the emotional, and asserting that the sensual can be experienced unmediated by either thought or emotion. Sontag's understanding of aesthetics is related, therefore, to her understanding of illness and how best to describe it. But, such an understanding, as Sontag presents it, must leave out the affective voices of patients who use metaphors to empower them to challenge the conventional medical narratives of illness that emphasize the patient's passivity. And yet, via affective misreadings, her work seems to have served as an inspiration for many ill people to write—metaphorically, more often than not—about their own experiences of illness. Furthermore, Sontag's work has most *effectively* de-mythologized disease simply by challenging taken-for-granted representations of illness rather than by successfully purifying the experience of illness of metaphor.

I have left out a crucial aspect of the book—*the* crucial aspect, perhaps—which is Sontag's motivation for writing *Illness as Metaphor.* Although she doesn't give readers any hint of it in the actual text of *Illness as Metaphor,* there is, paradoxically, a personal story of illness behind her desire to depersonalize the experience of illness. Sontag herself was diagnosed with and treated for cancer in the two years before she wrote her treatise on nineteenth- and twentieth-century representations of illness. By not mentioning this fact until now, I do not mean to imply that Sontag was in the closet about her cancer,[11] I simply wanted to highlight the fact that her desire to describe the possibility of having a *pure experience of illness* required that she *depersonalize* her own relationship to illness in her writing about illness.

On January 30, 1978, the same week that the first part of what would become *Illness as Metaphor* was published in the *New York Review of Books*, Sontag was interviewed in the *New York Times*, and the (unnamed) interviewer states, "she makes a point of openness about her illness" (A16). Moreover, she herself admits that her own first responses to her diagnosis were not in the form of an idea at all; rather, what she experienced in this moment was: "Panic. Animal terror. I found myself doing very primitive sorts of things, like sleeping with the light on the first couple of months. I was afraid of the dark. You really do feel as though you're looking into that black hole" (A16). In this statement, Sontag attempts to describe the ways in which the person who is ill experiences illness. She attempts, that is, to give her own terrifying experience form in language, and this terrifying experience is given form, at least initially, through metaphor. She *uses* metaphor to explain how illness makes her feel: she is "looking into that black hole" of her most primitive fears. But, it is important to reiterate, in the text of *Illness as Metaphor,* as it was published in the *New York Review of Books* and in its slightly revised book form, Sontag chooses not to reveal this particular affective history.[12]

Sontag's refusal of the affective, however, leads to some rather ironic readings in at least two of the reviews of the book version of *Illness as Metaphor.* In a review for the *New York Times,* John Leonard euphemizes Sontag's cancer as "[h]er own widely publicized health problems" (1978, C19). By speaking euphemistically about cancer, Leonard contributes to the stigmatizing ways in which cancer is perceived, which is, of course, precisely what *Illness as Metaphor* argues so effectively against. Despite his oblique reference, Leonard nonetheless believes that Sontag's health problems "doubtless account for the tone and content of *Illness as Metaphor,* but they also probably account for its lucidity. It is burned clean of mannerism and of glibness" (C19). Leonard's mostly positive review of *Illness as Metaphor* ends with an acknowledgement of both the usefulness of metaphors, in particular as "our way of thinking about death," as well as the necessity for someone, like Sontag, to be a "critic of metaphors" (C19). Her credentials to carry out such a task, it seems, are her experiences both as an intellectual—someone who does things with ideas—and as a person with cancer.

In contrast to Leonard, Denis Donoghue, in the *New York Times Book Review,* is frankly a bit baffled by what he perceives as Sontag's anger throughout *Illness as Metaphor.* From his comfortable position outside the kingdom of the ill, he offers a reasoned and reasonable—that is, *not angry*—assessment of Sontag's concerns in *Illness as Metaphor:* "If a doctor gave me a psychological stereotype instead of a cure or an alleviation, I'd demand my money back. If doctors have nothing better to say than that you have cancer because you are the type of person to get cancer, then indeed they should keep quiet. But because they don't know what causes cancer, their offense is venial if they hazard a guess" (1978, 9). The chasm between doctor and patient in relation to power/knowledge does not worry Donoghue. The passive patient role is not something Donoghue can imagine; he has absolutely no doubt that he would not become passive in that position, but would remain a person—an agent—with countless options and the ability to make demands. Despite his own gesture to the personal to dispute Sontag's picture of the doctor-patient relationship, Donoghue suspects that *Illness as Metaphor* "is a deeply personal book pretending for the sake of decency to be a thesis" (27). A personal book, Donoghue seems to say, can only pretend "for the sake of decency" to have a thesis. Finally, Donoghue is also not very concerned about the "sinister mythology of cancer" because he cannot believe that it "will persist after the causes of the disease are known and a successful treatment is produced" (9). This is precisely Sontag's point, though she recognizes as Donoghue does not, that cancer will not be the last disease with a sinister mythology. And, in just a few short years, another disease and its sinister mythology would appear, and this disease becomes further evidence for Sontag's argument.

Ten years after the publication of *Illness as Metaphor,* Sontag returned to her task of de-mythologizing disease; this time her critique was leveled against the rampant metaphorization of the newly discovered and sufficiently mysterious AIDS virus. What is remarkable about *AIDS and Its Metaphors* (1988) is not so much her delineation of the damaging metaphors attached to HIV/AIDS,[13] but her return to, and, to some extent, rewriting of, *Illness as Metaphor.* For someone "against interpretation," Sontag, surprisingly, offers her own interpretation of her earlier work. We learn or are reminded that Sontag was motivated to write *Illness as Metaphor* not only as a continuation of her work on representation in literature, visual art, and photography, but also because of her own experience with breast cancer. "Twelve years ago," Sontag writes, "when I became a cancer patient, what particularly enraged me—and distracted me from my own terror and despair at my doctor's gloomy prognosis—was seeing how much the very reputation of this illness added to the suffering of those who have it" (1988, 12). Why did Sontag's own feelings of terror and despair not appear in the text of *Illness as Metaphor*? Because, Sontag writes ten years later, "I didn't think it would be useful—and I wanted to be useful—to tell yet one more story in the first person of how someone learned that she or he had cancer, wept, struggled, was comforted, suffered, took courage . . . though mine was also that story. A narrative, it seemed to me, would be less useful than an idea" (13; ellipses in original). This is a somewhat strange, though revealing, opposition that Sontag sets up between a narrative and an idea. Presumably, she means specifically a *personal* (or, perhaps, affective) narrative, because, of course, even her idea must be given narrative form for it to be useful. In *AIDS and Its Metaphors*, Sontag wonders within the text itself what is most useful in the face of illness, and, in particular, what is most useful for the person who is ill and others like her.

In *The Cancer Journals*, Lorde too is concerned with the usefulness—the effectiveness—of her work. However, what Lorde will "share for use" is not an idea—though her axiom about "the transformation of silence into language and action" is, of course, as much an idea as anything else[14]—but a personal narrative that she hopes will also have political effects beyond her in time and space. Sontag, on the other hand, at least at the time she wrote *Illness as Metaphor*, believed that a choice must be made between a personal narrative and an idea. Only later, when writing not about an illness she has but about an illness among people she knows and loves, will she feel a need to clarify and put into print the personal motivation that fueled the writing of *Illness as Metaphor*. And so, in *AIDS and Its Metaphors*, Sontag admits she wrote *Illness as Metaphor* "spurred by evangelical zeal as well as anxiety about how much time I had left to do any living or writing in" (13), which is a concern Lorde will also articulate. But, Sontag explains, "The purpose of my book was to calm the imagination, not to incite it"(14), and, here, she parts company with both Lorde and Sedgwick, who as poets seek, precisely, to incite the imagination. Lorde and Sedgwick each have a poet's faith that metaphor can help not just hurt—that it may be, in fact, *useful*—for those who are living with cancer.

In *AIDS and Its Metaphors*, Sontag feels it is necessary to assert that *Illness as Metaphor* is a form of effective history; she calls her earlier work "an exhortation" to others to "[g]et the doctors to tell you the truth; be an informed, active patient; find yourself good treatment, because good treatment does exist (amid the widespread ineptitude)" (15). Yet, nowhere in *Illness as Metaphor* does Sontag make such statements, even obliquely. So, why rewrite her earlier work? Why supplement an idea with a personal story of suffering (though brief) and a political exhortation (though belated)? The answer, it seems to me, has something to do with both AIDS and metaphors. What Sontag experiences again in the face of HIV/AIDS is the panic and animal terror that her own diagnosis of breast cancer had brought with it over ten years before. In opening her work on AIDS with a return to her work before AIDS, Sontag seems to doubt the faith she had proclaimed and yet still wants to proclaim: that an idea alone can alleviate suffering, that an idea can illuminate once and for all the black hole of our most primitive fears. With HIV/AIDS the black hole returns, and although she faces it once more with an idea, there seems to be less certainty that an idea alone is enough. The affective arises in Sontag's narrative despite her attempts to repress it through the control that she believes her de-metaphorizing idea affords her.

In contrasting Sontag's illness narrative with Sedgwick's later narrative, I wonder if the sequence of illness events has something to do with their differing rhetorical approaches. In some respects, Sontag's work begins with the personal experience of illness, and yet she is convinced that bringing in this personal experience would undermine the effectiveness of her argument (and Donoghue's review of her work seems to bear this out, though it also shows the ways in which, nonetheless, the personal and affective was read into her work at the time of its publication, just as the metaphorical has been read into it since). Only when her personal experience of illness becomes, paradoxically, also about the personal experiences of illness *of others* does Sontag sense it might be necessary and effective to tell a piece of that personal story, at least as an introduction to her idea now applied to a new illness event. Sedgwick's relationship to the same two illness events experienced by Sontag (breast cancer and AIDS) is reversed. She had experienced the illness of others and written about AIDS before she was diagnosed with and would write about breast cancer. Her political, theoretical, and performative response to AIDS comes before her own experience with breast cancer, but also her ideas about illness emerge out of the affective associated with both AIDS and breast cancer rather than in opposition to it, as Sontag's do.

Before I discuss the ways in which Lorde and Sedgwick offer different forms of treatment in the face of breast cancer, I want to turn briefly to a critique by D.A. Miller of Sontag that reveals the ways in which Sontag's attempt to reduce the affective experience that surrounds illness to an idea diminishes the effectiveness of her argument. Although he is somewhat pleasantly surprised by Sontag's own rereading of *Illness as Metaphor* and the personal revelations with which she opens *AIDS and Its Metaphors*, Miller nonetheless offers a scathing reproach of what he calls Sontag's "urbanity," by which he means her over-intellectualizing, her excessive detachment from any affective response to the crisis of HIV/AIDS, and her focus on exclusively literary as opposed to ethical or political questions. According to Miller, the problem is not Sontag's "'views' on AIDS . . . so much as in the *attitude of her writing*," that is, "the unexamined and . . . largely unconscious complex of positionings, protocols, and poses that determine her deployment of language" (1993, 213; emphasis in original). Miller's reading of Sontag is important because it points directly to the ways in which Sontag's work itself belies the claims she makes for it.[15] Miller recognizes, moreover, that while it may be necessary to oppose certain damaging metaphors—like those associated with the "cancer personality"—it is also possible, and even necessary, to employ metaphor as resistance. He challenges Sontag's concluding "modest proposal" to retire the military imagery attached to illness, and suggests such a blanket retirement of military metaphors is disingenuous at best. According to Miller,

> [Sontag] forgets how well one such military metaphor—the one conveyed in the word "polemic" (15) (from the Greek *polemos* [war])—served her as a cancer patient, beset by debilitating myths of "responsibility" and "predisposition." She also overlooks how vital another such metaphor—the one conveyed in the word militancy (from the Latin *miles* [soldier])—is proving to people with AIDS and to the AIDS activism of which they stand at the center. (219)

To the terms "polemic" as a form of rhetoric and "militancy" as a form of praxis, we might add the terms "warrior" (as opposed to "victim") and "queer" as modes of subjectification and ways of doing illness and politicizing patienthood that Lorde and Sedgwick articulate in their illness narratives.

Doing Things with Experience

In her "biomythography," *Zami: A New Spelling of My Name* (1982), Audre Lorde describes a job she had in her early twenties in Stamford, Connecticut, reading crystals with a commercial X-ray machine at Keystone Electronics. Attuned to the race, class, and gender politics of work and health, Lorde explains how she and her co-workers came to work at Keystone Electronics:

> Most local people would not work under such conditions, so the cutting crew was composed of Puerto Ricans who were recruited in New York City and who commuted every morning up to Stamford on company-paid tickets. Women read the crystals on a variety of X-ray machines, or washed the thousands and thousands of crystals processed daily in huge vats of carbon tetrachloride.
>
> All the help in the plant, with the exception of the foreman and forewomen, were Black or Puerto Rican, and all the women were local, from the Stamford area.
>
> Nobody mentioned that carbon tet destroys the liver and causes cancer of the kidneys. Nobody mentioned that the X-ray machines, when used unshielded, delivered doses of constant low radiation far in excess of what was considered safe even in those days. Keystone Electronics hired Black women and didn't fire them after three weeks. We even got to join the union. (126)

The black and Puerto Rican women reading the crystals with an X-ray machine are provided with safety hoods to prevent the harmful X-rays from coming in contact with their bodies. But, flipping the hood and shielding one's body means losing valuable seconds, and that little bit of extra money. As this scene reveals, for Lorde, cancer has everything to do with being a poor, black woman in the United States, just as the countless tuberculosis deaths witnessed by Swan at Sioux Sanatorium had everything to do with being poor and Native American. Lorde doesn't discuss her cancer in *Zami*, which only recounts her childhood and young adulthood. Nonetheless, one cannot read the story told in *Zami* as isolated from the story *The Cancer Journals* (1980) tells before it. Lorde's cancer hovers over *Zami* and leaves its trace on that narrative, just as the carbon tetrachloride and radiation leave their traces in the cells of her body. The X-ray machine may allow her to *read* the crystal, but at the time she is blind to the real, which she encounters only too late: the real of cancer, which cannot be read, except as a symptom, a future symptom. While Lorde and her coworkers do not flip the hood to protect themselves, they are screened, nonetheless, by social circumstances and class and race positionings that prevent them from seeing a possible future—a future of illness, or, perhaps, a future free of illness—in the present moment. Lorde's rendering of her past experience reading crystals (because that was the only job a young, black woman could get in the United States in the 1950s) is haunted by the future cancer that she faces as she writes her "biomythography."[16]

During the same period in which Lorde was working on *Zami*, she was also undergoing treatment for breast cancer. On November 19, 1979, Lorde wrote:

> I want to write rage but all that comes is sadness. We have been sad long enough to make this earth either weep or grow fertile. I am an anachronism, a sport, like the bee that was never meant to fly. Science said so. I am not supposed to exist. I carry death around in my body like a condemnation. But I do live. The bee flies. There must be some way to integrate death into living, neither ignoring it nor giving in to it. (1980, 13)

Lorde's work is an early, clarion call to scrutinize the meanings of breast cancer and to "integrate this crisis into useful strengths for change" (10).[17] She asserts the need for cultural critique as well as and in combination with political action as a necessary alternative to the more usual responses to breast cancer at that time: prosthesis, silence, and invisibility. For autobiography scholar Thomas Couser, Lorde's work is particularly powerful because it "shifts back and forth between the proximate and the distant, between the emotional and the intellectual, as Lorde struggles to bring all her resources to bear on a new and frightening challenge" (1997, 50–51). "The strength of the book," Couser continues, "is in its inclusion of both her private responses—cries of pain and outrage—and her political analysis—seasoned, reasoned discourse" (51). What gives *The Cancer Journals* its urgency is its integration of the affective with the effective.

The Cancer Journals, like all of Lorde's work, is both past- and future-directed. "My work is a part of a continuum of women's work" (17), Lorde explains, acknowledging the inspiration of earlier women's writing and social activism. Her approach to her cancer is continuous with the transformative politics advocated by the women's health movement in the 1970s.[18] *The Cancer Journals* is connected to a genealogy of health feminism, whose line of descent also includes Ehrenreich and English's *For Her Own Good* (1978), Claudia Dreifus's *Seizing Our Bodies* (1977), Rose Kushner's *Why Me? What Every Woman Should Know About Breast Cancer to Save Her Life* (1975), the Boston Women's Health Collective's *Our Bodies, Ourselves* (1973), and Ellen Frankfort's *Vaginal Politics* (1972), all of which attempted to provide the necessary knowledge to make the patient her own health expert.

As the title of Dreifus's edited volume suggests, one of the main goals of this movement was that women seize control of their own bodies from established institutions, such as medicine and the law, as a means to liberation. Health feminism was concerned with women's often negative experience of medicine, and in particular gynecology. The "radical, anarchic" movement distrusted organized medicine, and offered an early critique of the medicalization of supposedly "natural" experiences, which resulted in, for example, the transformation of childbirth into an increasingly technological experience in the hands of usually male obstetricians (Dreifus 1977, xxiv–xxv). The women's health movement established both consciousness-raising groups, in which "expert" medical knowledge could be shared beyond the small coterie of medical professionals, and self-help clinics, which encouraged patient involvement in diagnosis and treatment. As Dreifus explains, though they "often employ staff physicians," these clinics, which would become the Feminist Women's Health Centers, "uniformly have women, often nonprofessionals, doing the bulk of the medical service; their emphasis, always, is on patient involvement in diagnosis and treatment, on deprofessionalization of services" (xxviii).

Just as importantly, and also as a means to raise consciousness about the politics of women's bodies, the women's health movement used direct action tactics to disrupt the specific institutions that affected women's experiences of their bodies. For example, a group of health feminists invaded the Senate hearings on the safety of oral contraceptives (Dreifus 1977, xxix). As Judith Coburn explained in 1970, "We went to the hearings to protest the alliance of drug companies, population control experts, and the government, an alliance we believe has given the Pill a clean bill of health at women's expense" (2000, 136). Although AIDS activists would demand something somewhat different from the drug companies and the government than those women who disrupted the Senate, the notion that one might challenge specific government institutions making health policy, like the Food and Drug Administration, emerged at this time.[19] The concern with questionable alliances that affect women's health also undergirds Lorde's critique of what she calls Cancer Inc. and Ehrenreich's later critique of the Cancer Industrial Complex.

Lorde's work takes these feminist practices and challenges to expert medical knowledges and applies them to the specific situation of breast cancer, and by doing so also prophetically[20] anticipates the AIDS activism of the 1980s[21] and the breast cancer activism of the 1990s, as well as the boom during this same time in personal narratives describing the experience of illness from the patient's perspective. While understanding that fear—of cancer, of death—"is an appropriate response to a real situation" that she must "learn to work through" (14), Lorde rejects the medical and lay view of the patient as passive. *The Cancer Journals*, then, is an affective history—with fear worked through rather than denied or banished—and an effective history—with the patient as active and articulate and seeking knowledge for herself and passing it on to others in her illness narrative. Or, as Lorde herself states succinctly, "I am not only a casualty, I am also a warrior" (21).

It is in this spirit that Lorde rejects prosthesis and reconstruction because she refuses to carry on as though her body remained unchanged by cancer and its treatments; she refuses to hide her scars and to pass as a two-breasted woman. She refuses, finally, to cover up her difference:

> [I]t is that very difference which I wish to affirm, because I have lived it, and survived it, and wish to share that strength with other women. If we are to translate the silence surrounding breast cancer into language and action against this scourge, then the first step is that women with mastectomies must become visible to each other. (61)

It is not that Lorde doesn't speak of loss; rather, her new sense of self and her political and personal voice emerges, precisely, out of loss. For Lorde, after breast cancer, after mastectomy, the self cannot be made the same as it was before breast cancer and mastectomy. There is no return. Her hope, then, is that by giving voice to loss, she might "share it for use, that the pain not be wasted" (16).

The task for Lorde is more than her own survival; she must also make her survival useful to others through writing, teaching, and activism. Survival is not something you do just once and then move on from. Lorde explains that, "[g]rowing up Fat Black Female and almost blind in america requires so much surviving that you have to learn from it or die" (40). Lorde's own survival is forever linked to those others who did not survive, whose names she has "tattooed upon [her] heart" (40), as a reminder—a body memory—that she carries within her.[22] Because it involves carrying forward the story of others, Lorde's struggle is ongoing. She emphasizes this *ongoingness* by summarizing the lessons she learns from her experiences of breast cancer as a series of questions—not simply as answers:

> How do I provide myself with the best physical and psychic nourishment to repair past, and to minimize future damage to my body? How do I give voice to my quests so that other women can take what they need from my experiences? How do my experiences with cancer fit into a larger tapestry of my work as a Black woman, into the history of all women? And most of all, how do I fight the despair born of fear and anger and powerlessness which is my greatest internal enemy? (16–17)

The politicization of patienthood is not about being given new answers to old questions; rather, as Lorde shows, it is about opening up a space for questioning itself. Illness enacts a transformation of the questions one can and must ask, and previously unimagined questions present themselves. For Lorde, patienthood is not at all what it was for MacDonald, who was expected not to read, write, talk, laugh, sing, reach. Although MacDonald does not explicitly mention not questioning in her discussion of the tuberculosis routine, it is clear that patients following the routine must, especially, not question. Indeed, if a patient must lie still and not talk, there is little opportunity to question the routine, or the medical practitioners who attempt to instill it as a way of life. Although Swan's oral history told to St. Pierre contains a ghostly trace of all those Native Americans who did not survive, the story *Madonna Swan* tells does not include a political critique of her treatment—or lack thereof—at the hands of white medicine. And, while Sontag's demythologizing does question the ways in which illness is represented, her refusal "to incite the imagination," means her questioning is somewhat limited in the form it can take. She must rely on ideas that are already available, not thoughts as yet unthought, still to be invented.

Although Lorde recognizes that the loss of a breast is a small sacrifice she must make in order to survive, she also knows the symbolic meanings attached to women's breasts, and recognizes that such a loss brings up at the same time the potential loss of a sense of self. Lorde understands what Iris Marion Young asserts in her essay on the phenomenology of "breasted experience": that a woman's breasts are "entwined with her sense of herself" (1990, 189).[23] Breasts symbolize woman as, paradoxically, both mother and sexual being; they are, "a scandal," according to Young, "because they shatter the border between motherhood and sexuality" (199). Thus, the loss of a breast highlights a symbolic loss of the capacity to mother, as well as the capacity to desire and be desirable. As Couser notes, then, breast cancer is dreaded not only because it may take a woman's life, but also because "it may destroy her womanliness" (1997, 46). And, in an essay entitled "Refugees in a Strange Country," Carole Colbourn, a doctor who has breast cancer, offers this "reasoned response" to a male surgeon who wonders, "What's so special about breasts anyway?":

> Answer 1. Ask any straight man. If he was being honest he would tell you it is a stupid question.
>
> Answer 2. Breast cancer kills children's mothers. (1996, 125)

Although Colbourn follows these first two responses with further information about the frequency of breast cancer in the West, its deadliness, and the scientific uncertainty surrounding its cause and possible treatments, she makes clear why breasts really matter (in patriarchal culture, one might add, though Colbourn herself, tellingly, does not): as objects to be desired by heterosexual men and as symbols of motherhood.

In contrast to Colbourn and others who believe breasts matter only as symbols of heterosexuality and motherhood, Lorde offers a black, lesbian, feminist analysis of breast cancer that critiques the notion of women as seen only as mothers or sexual objects, and of femininity as defined by appearance. Although she knows that in patriarchal culture the breast often stands in for the whole woman, she recognizes as well that a woman's being-in-the-world cannot be held by or contained in the breasts and their intact appearance. Lorde's critique of prosthesis and reconstruction, then, is framed by a feminist critique of "society's stereotype of women, that we are only what we look or appear" (1980, 57). Lorde problematizes the "false and dangerous nostalgia" (56) for a return to the way one was before breast cancer:

> Any woman who has had a breast removed because of cancer knows she does not feel the same. But we are allowed no psychic time or space to examine

> what our true feelings are, to make them our own. With quick cosmetic reassurance, we are told that our feelings are not important, our appearance is all, the sum total of self. (57)

In *The Cancer Journals*, Lorde provides narrative alternatives to prosthesis and breast reconstruction; the self is reconstructed—constructed anew not as before—through the practice of writing.[24]

Lorde explicitly links writing with her experience of mastectomy and her personal and political desire not to wear a prosthesis. She tells her reader, "On the day after the stitches came out and I got so furious with the nurse who told me I was bad for the morale of the office because I did not wear a prosthesis, I wrote in my journal" (52). The writing becomes a substitute for prosthesis, a response that, while not necessarily *good for morale*, is decidedly *ethical*. On that day, Lorde writes about writing, and, in particular, the impossibility of writing the whole story, the whole truth of her experience:

> There is so much I have not said in the past few days, that can only be lived now—the act of writing seems impossible to me sometimes, the space of time for the words to form or be written is long enough for the situation to totally alter, leaving you liar or at search once again for the truth. (52)

And yet, her writing itself is evidence of her need to continue writing as a practice of the self. For Lorde, writing about that which is being lived now necessarily fails; we can never find the words—the right words—nor the time—the right time—to fix the present. There is always a gap between the words and the thing itself: be it the life, the present moment, the truth, and/or the real of the experience of illness. But, for Lorde, what matters is not the fixing of a truth of the self in language, but the ethical practice of searching for such a truth in language, the conjunction between language and action as forms of doing.[25] When people recall Lorde's well-known dictum about the transformation of silence into language and action, they often emphasize a sequential relationship between language and action; that is, action follows language in a kind of politics of naming. It seems to me, however, that Lorde calls for language and action simultaneously; that is, *language as action and action as language*. To paraphrase Foucault, language is in itself an action—a perilous act (1977b, 5) that brings a new world into being.

In her work, Lorde articulates a form of politicized patienthood that seeks to transform society's normalizing functions in ways that are more radical than those Sontag articulates. Lorde offers a political critique of the regimes of health that takes into account the normative categories of gender, race, and class, and the ways that these intersect with the normative category of health. So, while Lorde encourages potentially painful *self scrutiny* and evaluation of one's life in the face of death, and recognizes that such scrutiny can be a "rewarding and strengthening journey . . . towards a deeper self" (58), she also encourages a concomitant *social and cultural scrutiny* into the norms of femininity and heterosexuality as well as into the hegemony of the profit economy and its potential causal relationship to cancer. Whereas Sontag's rhetorical analysis seeks to purify representations of illness of potentially debilitating metaphors, Lorde's materialist analysis seeks to uncover institutional and structural factors that prevent women (and men) from responding effectively to the experience and event of cancer. To this end, Lorde takes on the "american medical establishment" and its public voice, the American Cancer Society:

> Any information about the prevention or treatment of breast cancer which might possibly threaten the vested interests of the american medical establishment is difficult to acquire in this country. Only through continuing scrutiny of various non-mainstream sources of information, such as alternative and women's presses, can a picture of new possibilities for prevention and treatment of breast cancer emerge. (71)

The bottom line, according to Lorde's materialist analysis, is that "there is no profit in the prevention of cancer; there is only profit in the treatment of cancer" (71). Lorde is concerned, then, not only with personal silencing as a result of fear and shame, but also with the institutional silencing of other knowledges and practices of health by what she calls "Cancer Inc." (62). "Every woman has a militant responsibility," she asserts, "to involve herself actively with her own health" (73).

This "militant responsibility," however, does not mean a woman is responsible for her own breast cancer, nor is she alone responsible for getting better. Lorde emphasizes the need to be a "fighter resisting rather than . . . a passive victim suffering," and that it is "physically important for me to be loving my life rather than to be mourning my breast" (73); but this affirmation of self is quite different from "the superficial farce of 'looking on the bright side of things'" (74). Lorde exposes the depoliticizing effect of "blame-the-victim thinking," and, like Sontag, forcefully dismisses the idea that there might be a "cancer personality." This "blame-the-victim syndrome," Lorde insists,

> does nothing to encourage the mobilization of our psychic defenses against the very real forms of death which surround us. It is easier to demand happiness than to clean up the environment. The acceptance of illusion and appearance as reality is another symptom of this same refusal to examine

> the realities of our lives. Let us seek "joy" rather than real food and clean air and a saner future on a liveable earth! As if happiness alone can protect us from the results of profit-madness. (74)

According to Lorde, health is first and foremost a political issue; it is not an effect of or a means to personal happiness. What is more important than achieving surface happiness is "recognizing the deep and fundamental unhappiness with which we are surrounded, at the same time as we fight to keep from being submerged by it" (75). Breast cancer cannot, finally, be understood as separate from either the social or physical environment, nor from sexism, racism, and classism.[26] Lorde's experiences as a black, feminist, lesbian, mother, warrior, and poet affect her art of doing illness. Her particular standpoint allows her to see the social structures that shape illness, and contributes to her skepticism of the simplistic belief that joy and happiness can come through personal transformation without a concomitant social and political transformation. And although her politics are lesbian and feminist, and so perhaps not yet queer, like Sedgwick after her, Lorde is interested in what we might do politically with affects, and not just the positive affects, but the negative ones, including unhappiness, anger, and shame.

How to Do Things with Shame

Along with Judith Butler (1990, 1993a, 1993b, 1997a) and Jacques Derrida (1982), Eve Sedgwick has been at the forefront of a move to make use of J. L. Austin's idea of performativity to contest essentialist notions of identity in general and of gender and sexuality in particular. In their introduction to the collection *Performativity and Performance* (Parker and Sedgwick 1995), the editors begin with two questions that emerge out of their reading of J. L. Austin's *How to Do Things with Words* (1961)[27]: "When is saying something doing something? And how is saying something doing something?" (1). A performative utterance, according to Austin, is that which enacts something in the moment it is spoken. One example of a performative that Austin gives, and Butler, Parker and Sedgwick, and others have elaborated on quite extensively, is the statement "I do [take this man or woman to be my lawfully wedded husband or wife]." In saying "I do" one becomes something one wasn't in the moment before the saying: husband or wife (presuming, of course, that the social context for this performance is, in Austin's terminology, "felicitous"). What many commentators on Austin have noted, and which Austin himself makes much of, is that many things can go wrong in the performance of a performative making that performative "infelicitous" or "unhappy"; in other words, as Austin also asserts, a performative can become "ill" (Parker and Sedgwick, 3; Austin, 18–19). In their reading of Austin, Parker and Sedgwick place the possibility—or, indeed, the inevitability—of a performative becoming ill at the center of the notion of the performative; they note, "illness [is] . . . understood here as intrinsic to and thus constitutive of the structure of performatives" (3). Thus, according to Parker and Sedgwick, "a performative utterance is one, as it were, that always may get sick" (3). In their discussion, Parker and Sedgwick then move from the possibility of an "ill" performative to the somewhat analogous notions of a "perverse" or "queer" performative in order to discuss performativity in relation to the example of "the Pentagon's 1993 'don't ask, don't tell, don't pursue' policy on lesbians and gay men in the U.S. military" (5). I want to move the other way—back from "queer" to "ill"—in order to make use of Sedgwick's concept of "queer performativity" to understand another mode for doing illness politically.

I will begin, then, by delineating what Sedgwick means by "queer" before determining the ways in which a queer performative might be useful in relation to the experience of patienthood, and putting that experience, in Lorde's words, into language and action. In her collection of essays entitled *Tendencies* (1993b), Sedgwick explains that "something about *queer* is inextinguishable," and defines "queer" as:

> a continuing moment, movement, motive—recurrent, eddying, *troublant*. The word "queer" itself means *across*—it comes from the Indo-European root *–twerkw*, which also yields the German *quer* (transverse), Latin *torquere* (to twist), English *athwart*. . . . The immemorial current that *queer* represents is antiseparatist as it is antiassimilationist. Keenly, it is relational, and strange. (xii)

Queer, for Sedgwick then, is a continuing movement across bodies and differences. It is at once relational—it perceives beings in relation—and strange—it doesn't attempt to make anyone's gender or sexuality "signifiy monolithically" (8).

In her essay "Queer Performativity" (1993a), Sedgwick discusses the plethora of recent uses of Judith Butler's notion of performativity, and worries that such uses are "a sadly premature domestication of a conceptual tool whose powers we really have barely yet to explore," because they all generally reach the same conclusion: that a particular performance is "really *parodic and subversive* (e.g. of gender essentialism) or just *uphold[s] the same status quo*" (15; emphasis in original). Sedgwick wants to go beyond these dichotomous good performative/bad performative formulations, and in an attempt to use the concept of performativity (from both Butler and Austin) more radically, she explores it as a means of "understanding the obliquities among *meaning, being,* and *doing*; not only around the examples of drag performance and (its derivative?) gendered self-presentation, but equally for such complex speech acts as coming out, for work around AIDS and other grave identity-implicating illnesses, and for the self-labeled, transversely but urgently representational placarded body of *demonstration*" (2; emphasis in the original). At this point in Sedgwick's work, the experience of illness, not the rhetorical notion of an ill performative, becomes a means by which one might imagine a more radical form of performativity, or indeed, a specifically queer form of performativity.

Although for Sedgwick the concept of queer emerged specifically out of her work on gender and sexuality, she recognizes how useful—personally, politically, and theoretically—it might be where AIDS as well as her own breast cancer are concerned to have those experiences of illness confront the theoretical and political models that had helped her "make sense of the world so far":

> The phenomenology of life-threatening illness; the performativity of a life threatened, relatively early on, by illness; the recent crystallization of a politics explicitly oriented around grave illness; exploring these connections *has* (at least for me it has) to mean hurling my energies outward to inhabit the very farthest of the loose ends where representation, identity, gender, sexuality, and the body can't be made to line up neatly together. (1993b, 13; emphasis in the original)

For Sedgwick, exploring the connections (and disconnections) between modes of being ill, the meanings attached to illness, and the politics surrounding illness (that is, between forms of meaning, being, and doing that surround the experience and event of illness), requires that she hurl her energies outward into new forms of representation and embodiment, rather than, as Sontag proposes, inward into a "pure, untranslatable, sensuous immediacy" (1966, 9). Sedgwick doesn't expect this confrontation to create a work that is "unified and clean," as Sontag aspires to. Rather, it is only in "mucking about in" the panic, terror, and other affects attached to the experience of illness that one might invent new forms of being patient, and new languages for representing that experience, as Lorde's work attempts to do as well.

For Sedgwick, one affect in particular—shame—will connect the two illness events—AIDS and breast cancer—that, although not always the explicit focus of her work, nonetheless haunt much of it.[28] In many respects, Sontag too is responding in her work to the shame that accompanies patienthood, but she responds by attempting to do away with this shame by arguing that illness has nothing to do with who a person is. Sedgwick believes, on the contrary, that illness does in fact have something to do with who one is, and she responds by attempting to do something else with the shame that is experienced along with illness. For her, as for Lorde, shame is an affect that produces and delineates identity, usually a stigmatized identity: "Shame, as opposed to guilt, is a bad feeling that does not attach to what one does, but to what one is" (1993a, 12). Shame is used to manage identity, and, moreover, though Sedgwick does not mention this specifically, it is used to manage desire as well.[29] At the same time, however, shame can be a "near-exhaustible source of transformational energy": that is, an "experimental, creative, performative force" (4). Shame is productive not only of normalizing identifications, but also of transgressive disidentifications. Sedgwick wants to make the shame that one experiences along with illness creative of new ways of being and doing. Sedgwick's project is similar, then, to Lorde's desire to make women who have undergone mastectomy visible to one another in order to create a community of one-breasted women, who together seek new knowledge about the causes of and treatments for breast cancer.

In her own illness narrative "White Glasses," which is the final essay in *Tendencies*, Sedgwick writes about both AIDS and breast cancer; in fact, she recognizes that there is a "dialectical epistemology between the two diseases" (15). This "dialectical epistemology" has emerged out of the history of the two illness events in the West. As Sedgwick notes, and as I argued earlier, the AIDS activism that emerged in the 1980s was influenced by the women's health movement of the 1970s, which encouraged women to become experts on their own bodies and to challenge their objectification within the institution of medicine with new knowledge and new de-institutionalized practices of health care. In turn, according to Sedgwick, "an activist politics of breast cancer, spearheaded by lesbians, seems in the last year or two to have been emerging based on the model of AIDS activism" (15).[30] This dialectical epistemology between AIDS and breast cancer is not only demonstrated through the activism surrounding these diseases, but also in terms of "the kinds of secret each has constituted" and "the kinds of *out*ness each has required and inspired" (15). That is, Sedgwick wonders how shame has operated on those who are ill with AIDS and breast cancer, and how that shame has been a source of new forms of outness. For Sedgwick, moreover, the two diseases have "made an intimate motive for me," a phrase that I believe reveals in its odd locution Sedgwick's definition of "queer" as movement across, as relational, and as strange. How can we create intimacy out of illness—across bodies and across differences—and how can that intimacy be a motive to create new forms of embodiment, representation, and politics?

"White Glasses" opens with a confession of sorts, about how Sedgwick in setting out to write the piece "got everything wrong" (1993b, 255). As she discovers, she got everything wrong simply because she thought the difference between health and illness and between living and dying was clearly demarcated, that those categories were, as Sontag might say, "unified and clean":

> When I decided to write "White Glasses" four months ago, I thought my friend Michael Lynch was dying and I thought I was healthy. Unreflecting, I formed my identity as the prospective writer of this piece around the obituary presumption that my own frame for speaking, the margin of my survival and exemption, was the clearest thing in the world. In fact it was totally opaque: Michael didn't die; I wasn't healthy: within the space of a couple of weeks, we were dealing with a breathtaking revival of Michael's energy, alertness, appetite—also with my unexpected diagnosis with breast cancer already metastasized to several lymph nodes. (255)

It is impossible, Sedgwick discovers, to make this narrative unified and clean; instead, she must get everything wrong, and yet still attempt to move across the binaries between health and illness, living and dying, between Michael's identity and her own identity.[31] This position of getting everything wrong doesn't paralyze Sedgwick with shame; rather, it allows her to create identifications across difference, and, even, "across the ontological crack between the living and the dead" (257). According to Sedgwick, it is "exciting that Michael is alive and full of beans today, sick as he is; I think it is exciting to both of us that I am; and in many ways it is full of stimulation and interest, even, to be ill and writing" (256). It is exciting to both Lynch and Sedgwick that she is what exactly? Sick? Alive and full of beans today? Sick *and yet* alive and full of beans today? Or, simply, that she is? Here Sedgwick leaves open rather than closes down (or purifies) the affective that permeates her own experience of illness, which is always also more than hers alone, as Lorde recognized in linking her story of survival to the story of all those before her who did not survive. Moreover, as Lorde also suggests, a crucial aspect of an effective and affective history is leaving open the possibilities that might emerge out of the relationship between being ill and writing.

Although Sontag's work is effective in challenging the normalizing judgments attached to the experience of illness, it is less effective in challenging the binary relationship between health and illness. She successfully critiques the troubling metaphors used to describe the experience of illness, but does not consider questions of how to live as a person who is ill, as Lorde and

Sedgwick do. In his late work on ethics, Foucault contrasts the notion of a fixed gay identity with the more open possibility of trying to define and develop a gay way of life or art of existence. In an interview for the French magazine *Gai Pied* that appeared in April 1981, Foucault describes how this might work: "Homosexuality is a historic occasion to reopen affective and relational virtualities, not so much through the intrinsic qualities of the homosexual but because the 'slantwise' position of the latter, as it were, the diagonal lines he can lay out in the social fabric allow these virtualities to come to light" (1997, 138). Sedgwick is concerned with how to live a queer life and an ill life, and the potentially intimate relationships between the two, and she recognizes that she is fortunate to be surrounded by models for how to live as a person who is ill. This is, in fact, one thing she has learned and continues to learn from Michael Lynch:[32]

> So much about how to be sick—how to occupy most truthfully and powerfully, and at the same time constantly to question and deconstruct, the sick role, the identity of the "person living with life-threatening disease"—had long been embodied in him, and performed by him, in ways which many of us, sick and well, have had reason to appreciate keenly. (1993b, 261)

Two of the many lessons that Sedgwick learns from Michael Lynch are also two examples of a queer performative: "Out, out" and "Include, include." These queer performatives induce persons living with life-threatening illnesses "to entrust as many people as one possibly can with one's actual body and its needs, one's stories about its fate, one's dreams and one's sources of information or hypothesis about disease, cure, consolation, denial, and the state or institutional violence that are also invested in one's illness" (261). Sedgwick believes that "transformative political work" can be done by making oneself "available to be identified with in the very grain of one's illness (which is to say, the grain of one's own intellectual, emotional, bodily self as refracted through illness and as resistant to it)" (261). Yes, ideas are useful in the face of life-threatening disease, but so are stories, dreams, and hypotheses about the panic, terror, shame, and the institutional violence that we endure and resist. To queer the experience of patienthood is to "include, include" not to "purify, purify."

"Breast-Cancer Culture" and the Discourse of Responsibility

Having discussed the multiple ways that patienthood gets politicized in several breast cancer narratives, I want to conclude this chapter with the covering over of the discourse of politicization with another discourse, which might be called the "discourse of personal responsibility." I take this term from Jackie Stacey, whose book *Teratologies* is a "cultural study of cancer" that seeks to combine, as Sedgwick does in "White Glasses," "different modes or 'registers' of writing: personal, political, and theoretical." "In particular," Stacey continues, "I want to use my experience as a series of textual 'rhetorics of the self' that connect competing forms of knowledge, and indeed allow me to interrogate the very formations of these different knowledges" (1997, 24). One such "rhetorics of the self" is the discourse of responsibility that comes out of a therapeutic culture that has in many instances replaced a political culture.

In order to delineate this discourse, Stacy describes a visit to a therapist who asks her if she knows why she has cancer. The therapist wants Stacey to find an emotional cause (most likely somewhere in the distant past) with present physical effects. When she refuses to engage in his teleological game of emotional cause and physical effect, he wonders if she is "afraid of self-exploration" (202). Stacey notes that her personal encounter with the discourse of responsibility emerges within a larger political context in which the activism of the 1970s that emphasized the patient as her own expert (the same activism that inspired and influenced Lorde and Sedgwick) was transformed in the 1980s and 1990s into the New Right philosophies that promote what Stacey calls "hyperindividualism": "the individual as the supreme authority; the individual as the agent of his fate; the individual as culpable in the face of failure" (211). Stacey also maintains that the turn towards alternative medicine in the 1980s, and its emphasis on "self-health" is a key component of the discourse of responsibility. She acknowledges that self-health discourses can initially be empowering, especially in relation to the "alienating" experience of modern medicine, and she readily admits having incorporated some self-health practices into her own treatment regimen. Nonetheless, she agrees with Rosalind Coward that the increased interest in and use of alternative therapies has not amounted to a "thorough *political* critique of existing health service practices" (Coward 1989, 10). According to Stacey, "these self-health discourses (like their Thatcherite counterparts) lose sight of any social, economic or environmental forces which affect health and offer instead an entirely individualistic world view which posits the patient and her/his willpower at the centre of a manipulable universe" (1997, 221). I will return to Stacey's *Teratologies* again in the following chapter, but I mention it here to introduce the shift from the politicization to the depoliticization of patienthood, in particular in relation to the event of breast cancer in the United States at the turn of the twenty-first century.

In the 1990s, breast cancer activism became a movement in the United States, but the construction of patienthood in much of this movement is somewhat different from the construction that connects Sedgwick, Lorde, and Sontag with AIDS activism and women's health activism. In a cover story in the *New York Times Magazine* published in 1993, Susan Ferraro discusses the "anguished politics of breast cancer," citing the publication of *Dr. Susan Love's Breast Book* in 1990 and the formation of the National Breast Cancer Coalition in 1991 as key moments in the emergence of what she calls a "revolution." According to Ferraro, the breast cancer movement began in the 1970s when several "prominent women—coincidentally all of them figures in Republican Party politics—went public with the disease," including Shirley Temple Black in 1972, Betty Ford in 1974, and Happy Rockefeller in 1976 (1993, 58). Tellingly, she mentions Betty Rollin's *First, You Cry* (1976), but does not mention Lorde's *Cancer Journals*.

The women Ferraro interviews are portrayed as unlikely activists—an "odd sort of army" (27)—who show *surprising* political savvy. In contrast to AIDS activists, they are, Ferraro maintains, "(so far) less confrontational" (58), though she does mention WHAM (Women's Health Action and Mobilization) member Matuschka, who "makes art of her mastectomy with poster-size, one-breasted self-portraits that force people to see what cancer does" (58). More often though, according to Ferraro, these unlikely activists opt for coalition-building strategies and lobbying to influence those in government or the medical profession whose support might be beneficial to breast cancer survivors.

By the beginning of the twenty-first century, Ferraro's parenthetical "so far" would seem mostly wishful thinking as far as a confrontational breast cancer politics is concerned. In a scathing diagnosis of the breast cancer movement in the United States at the turn of the millennium, Barbara Ehrenreich asserts that something she calls "breast-cancer culture" threatens to replace breast cancer activism. When she discovers she has breast cancer, Ehrenreich admits that "even more unsettling than the growing prospect of major disease" is the noxious reality of "suffocation by the pink sticky sentiment" of breast cancer teddy bears and other kitsch that surrounds what she calls, signaling its professionalization not politicization, the "career of a breast-cancer patient" (2001, 43–44). She notes that this career "has been pretty well mapped out in advance" for her (44), and involves mostly the patient as infantilized consumer of cancer kitsch. Ehrenreich looks for continuities between the "spreading breast-cancer sisterhood" and the women's health movement, which she participated in in the 1970s and 1980s. She reminds her readers that the women's health movement "legitimized self-help and mutual support and encouraged women to network directly, sharing their stories, questioning the doctors, banding together" (47). She explains that, where breast cancer in particular was concerned, feminist health activists in the 1970s pushed for less radical and disfiguring treatment and ended the one-step surgery, in which a woman would have a mastectomy immediately following a biopsy that was discovered to be malignant.[33] By not waking her up to seek her input in her own treatment, the physicians performing this one-step surgery effectively silenced the patient and prevented her from participating in treatment decisions, not unlike the position MacDonald found herself in at The Pines in the 1930s.

Ehrenreich acknowledges that feminist health activism continues, in particular in the form of advocating for research into the environmental causes of breast cancer because determining causes will help the effort at prevention (47).[34] But, feminists and their modes of politicizing health are not a part of the dominant framework of the breast cancer movement, what Ehrenreich identifies as "mainstream breast-cancer culture" (48). Indeed, according to Ehrenreich, "[i]n the mainstream of breast-cancer culture, one finds very little anger, no mention of possible environmental causes, few complaints about the fact that, in all but the more advanced, metastasized cases, it is the 'treatments,' not the disease, that cause illness and pain" (48).

Recalling Lorde's "superficial farce of 'looking on the bright side of things,'" Ehrenreich decries what she sees as the "relentless brightsiding" of mainstream breast-cancer culture (49). According to this brightsiding ideology, breast cancer is not a dreadful disease that we must find ways to prevent, rather it is a "chance for creative self-transformation—a makeover opportunity, in fact" (49). To test the hold of the brightsiding ideology, Ehrenreich posts a comment on the Komen.org message board under the subject "angry." The responses to her comment are an almost universal condemnation of her anger, not a condemnation of the failure to prevent breast cancer or treat it effectively that has made her angry. Indeed, her respondents insist that anger is a wrong-headed—even dangerous—way to respond to cancer. Her test reveals that "cheerfulness is more or less mandatory, dissent a kind of treason" (50) in mainstream breast-cancer culture. Recalling Sedgwick's interest in the complex speech acts that occur in demonstrations, Ehrenreich wishes that women would do more demonstrating to get the government to invest more in finding the causes of cancer rather than racing for the cure, which focuses on early detection and treatment of cancer not its prevention.

In her critique of neoliberalism, Lisa Duggan identifies two key terms that are associated with neoliberal economic and cultural policies: privatization and personal responsibility.[35] According to Duggan, neoliberal rhetoric and practice "promotes the privatization of the costs of social reproduction, along with the care of human dependency needs, through *personal responsibility* exercised in the family and in civil society—thus shifting the costs from state agencies to individuals and households (2003, 14). We can see this promotion of privatization and personal responsibility in relation to breast cancer in the emphasis in mainstream breast cancer culture on individual awareness of risk factors, especially one's genetic predisposition and reproductive history, and early detection. The focus is on what the individual can and should do to determine her own risks for cancer, and her vigilance in monitoring her body so that she might detect her cancer early enough to be successfully treated. What is missing from this approach is a structural analysis that is concerned first and foremost with prevention of the disease in the first place. Ehrenreich exposes the specious logic of early detection, asserting that it is simply the portal "to treatments offering dubious protection and considerable collateral damage" (52).

She believes that the awareness and early detection mantras constitute little more than an "outbreak of mass delusion"—not just a breast-cancer *culture* but a breast-cancer *cult* – that "celebrat[es] survivorhood by downplaying mortality and promoting obedience to medical protocols known to have limited efficacy" (52). Like Lorde before her, Ehrenreich takes on the Cancer Industrial Complex, the plethora of multinational companies that create both disease, by releasing carcinogens into the environment, and cure, by offering "expensive, semi-toxic pharmaceutical treatments" (52). As Lorde also surmised, Ehrenreich believes the American Cancer Society is a key player in the Cancer Industrial Complex because it encourages the mass delusion that early detection is the solution to the problem of breast cancer, and also cynically avoids questions about possible environmental causes of cancer. The discourses and practices of the politicization of health, which seek to democratize expert knowledge and oppose personal responses to structural problems, have been covered over by neoliberal discourses and practices of personal responsibility.

Still, the politicized patient survives in the work of Ehrenreich and others,[36] which isn't to say that she—or the politicized patient more generally—is a "breast-cancer survivor." Ehrenreich abjures the term "survivor." She insists that in the breast-cancer cult the "mindless triumphalism of 'survivorhood' denigrates the dead and the dying" (53), because, according to this rhetoric, the dead and dying have failed: failed to survive and so failed to have cancer improve their lives. The breast-cancer cult's obsession with survivorhood is a far cry from Lorde's determination to find "some way to integrate death into living, neither ignoring it nor giving in to it" (1980, 13). Via this rhetoric of survivorhood, those who did not survive are somehow responsible for their own deaths, and we are back to the blame-the-victim logic that Sontag, Lorde, and Sedgwick all sought to eliminate through their discourses and practices of politicization. And yet, despite the louder rhetoric of the brightsiders, an echo of the politicized patient reverberates from Sontag to Lorde to Sedgwick to Ehrenreich, from the women's health movement to the AIDS movement to the breast cancer movement. Although I've presented the politicization of patienthood chronologically, the echo isn't unidirectional. It also reverberates from Ehrenreich back in time. In order to hear the echo that reverberates between individuals and across movements, we must treat breast cancer as both personal experience and sociopolitical event.

3.

Stories for and against the Self

Breast Cancer Narratives from the United States and Britain

Jackie Stacey begins *Teratologies* with the chapter "Heroes," in which she critiques the Western cultural narratives available to her for describing the experience of cancer, or, for that matter, any crisis of the self. According to Stacey, "In contemporary Western culture, we are encouraged to think of our lives as coherent stories of success, progress and movement. Loss and failure have their place but only as part of a broader picture of ascendance. The steady upward curve is the favoured contour" (1997, 9). With regards to illness, this need to move quickly beyond loss and failure, to show, indeed, that it has been left behind, means that the socially sanctioned illness narratives dwell not on loss and failure but on *the overcoming of loss and failure*. According to Stacey, the "fantasies of heroic recoveries and guaranteed survival" so common in narratives of cancer reveal one of the meanings of the word *teratologies*, that is, "the tales of monsters and marvels that pervade the popular imaginary of cancer subcultures" (10). Stacey wonders what is missing from this "crisis-rescue-recovery formulation," and she asks, "[w]hat remains untold in these heroic narratives? What does linearity exclude? What cannot be restored with closure? Where is the continued chaos and disorder in such accounts? Where is the forgotten pain?" (14–15). In other words, and a central question in *Treatments*, Where are the stories of failure? Where are the stories not of consolation or of a compensatory imaginary that screens the real, but of dissolution and even perhaps a desire for dissolution?

In a footnote, Stacey explains that her understanding of the "heroisation of the cancer patient and the progress narrative" comes from her own experience of the "cancer subcultures of 1990s Britain" that includes numerous influential books about living with cancer as well as alternative health practices like those associated with the Bristol Cancer Health Clinic (246–47, note 6).[1] Stacey acknowledges that she has only recently encountered, through the work of Arthur W. Frank, narratives that are counter to the "heroic cancer cultures of Britain" (247). Stacey, following Frank, asserts that there are "signs of a significant shift in the United States towards a different and less moralistic kind of storytelling," and that "perhaps similar changes may follow in the British context in due course" (247). My assessment of the British and American contexts differs from Stacey's. I think her work is part of a wider British resistance to cultural narratives of heroism, triumph, and the promise of self help (or "self-health" as Stacey calls it), which have emerged out of the British context in the 1980s and 1990s. I don't at all mean to say that it is simply a matter of Americans being into self-help and Britons being resistant to it. Like Stacey, I focus on larger "cultural narratives" that are revealed in illness narratives that are being written and read in a particular historical moment in the British and American contexts. And, while I think Stacey is right to identify cultural narratives of heroism as a generally Western phenomenon, I will diagram some differences that have emerged out of these different contexts that often get lumped together as "Anglo-American" or "Western."

In order to investigate such divergent British and American "cultural narratives" that inform recent writings about illness, I discuss in this chapter two breast cancer narratives—Sandra Butler and Barbara Rosenblum's *Cancer in Two Voices* (1991) and Ruth Picardie's *Before I Say Goodbye* (1998). In particular, I consider the ways in which each autobiography articulates a very different mode of being ill and doing and writing about illness, which I associate, rather schematically, with an American emphasis on the cultivation of an improved self, as exemplified by Butler and Rosenblum's work, and a British emphasis on the cultivation of an ironic self, as exemplified by Picardie's work. Before I consider the ways in which they are different, however, I first want to consider the ways in which they are similar. In the first part of this chapter, then, I look at the ways in which both narratives highlight the phenomenological transformation of the body in illness and the necessity of a concomitant transformation of narrative in writing about the embodied experience of illness.

Like the breast cancer narratives discussed in chapter 2, *Cancer in Two Voices* and *Before I Say Goodbye* scrutinize the phenomenological, psychological, and social meanings attached to the experience of breast cancer. They reveal the bodily transformations that breast cancer and its treatments enact, as well as the ways in which the self and others respond to such transformations. In illness the body becomes strange to the self, and both of these breast cancer narratives attempt to describe this estrangement of the self from the body. The experience of estrangement that illness enacts might be understood in terms of Freud's notion of the uncanny *(unheimlich)*, which he defines as something that is difficult to define. The uncanny is that which "arouses dread and creeping horror," but also, paradoxically, that which "leads back to something long known to us, once very familiar" (1953 [1919], 368, 369–70). As Freud understands it, something entirely known and fully perceived is not terrifying; rather, what is ambiguous and contradictory is the source of dread. Freud explains that the uncanny, or the "unhomely," in a translation of the German "*unheimlich*" that better reveals the ambiguity Freud wants to highlight, is the opposite of both meanings of the word "*heimlich*": that which is both "familiar and congenial" and which is also "concealed and kept out of sight," or domesticated (375). Thus, "the uncanny is in reality nothing new or foreign, but something familiar and old-established in the mind that has been estranged only by the process of repression" (394). Although Freud's definition is psychological ("something familiar and old-established *in the mind*"), his formulation of the ego from his earliest work is as an embodied ego, and the examples he presents in his 1919 essay of those things that are perceived as uncanny are all related to embodiment: dismembered limbs, a severed head, a hand cut off at the wrist, the experience of being buried alive, the idea of a "living doll," or the theme of the "double" (386–87, 397). Affective history, as I conceive it through the treatments of illness narratives presented in this book, attempts to negotiate this crisis of embodiment that the uncanny signals in order to consider the ways in which one must pay attention to what one has taken for granted, made familiar and congenial, and kept out of sight (from the self and others). Thus, in the first part of this chapter, I want simply to establish the ways in which the body, knowledge, and language are all defamiliarized in illness.[2]

Barbara Rosenblum, a white, working-class, lesbian, feminist sociologist, challenges, as Lorde does, the notion that the loss of a breast is a "diminishment of her femininity," and she finds the detailed discussions in her breast cancer support group about reconstructive surgery to have a "fetishistic quality" that covers over feelings of desperation and fear. Encountering her body in a mirror after mastectomy, Lorde admits that she "looked strange and uneven and peculiar" to herself, but she also recognizes that the "blush-pink nylon" prosthesis she has just tried on is even stranger than her new, one-breasted reality (1980, 44). Also like Lorde, Rosenblum understands that there is no possible return—either physically or emotionally—to her former embodied self; rather, what becomes constant for Rosenblum is, paradoxically, "liv[ing] in a body that keeps on changing" (1991, 136). This constantly changing body requires "slavish attention" and "an unnatural hypervigilance toward any and all sensations that occur within the landscape of the body" (136). There is no opportunity for taking the body for granted or forgetting it; it is unpredictable, and this unpredictability produces an ever-present terror and confusion. As Sedgwick discovered when she thought she was writing about Michael Lynch's illness, but turned out to be also writing about her own, the body as stable and invulnerable is a fantasy, but to lose this fantasy—to be faced with the everyday reality of instability and vulnerability—is to be forced to face the body's failures and to mourn its losses. As Rosenblum describes it: "I was thrown into a crisis of meaning. I could no longer assess and evaluate what sensations meant. I could no longer measure the intensity of sensations. I was no longer fluent in the language of my body, its signs and symbols, and I felt lost" (137).

When the knowledge of one's body and self that one has come to trust is no longer reliable, this also shatters one's faith in the possibility of meanings that are stable and can be shared. According to Rosenblum, "When you have cancer, the ground is pulled out from under you. Existence is problematic and anxious. You must look for new, stable ground" (137–38). But, this new, stable ground can now only be understood as, paradoxically, contingent and impermanent. Similarly, Lorde wonders how "to announce or preserve [her] new status as temporary upon this earth," before recognizing that this "new status" isn't new at all. "We have always been temporary," Lorde comes to understand, and yet it is only with breast cancer that she is able to underline this fact of always being temporary and to act on it, to share this knowledge—of vulnerability and contingency—for use (1980, 52). One must mourn, therefore, not only the losses of the body's capacities and the self's fantasy of coherence, but also the loss of a faith in those "previous systems of interpreting the body" and the world (Butler and Rosenblum 1991, 136). Rosenblum feels "hostage to the capriciousness of [her] body, a body that sabotages [her] sense of a continuous and taken for granted reality" (138), but, at the same time, this capriciousness, this sabotaging of all stable structures of meaning, forces Rosenblum to throw away the old structures and create new ones that are less permanent but no less essential to her beingness.

Picardie, like Rosenblum, understands breast cancer as an experience in which her body progressively becomes strange. But, for Picardie, the experience of her body as uncanny begins not with her diagnosis of cancer, but with her earlier experience of pregnancy. Picardie, along with many others (including Rose [1995] and Stacey [1997]), is conscious of the eerie similarities between pregnancy and cancer. For her, the fact that her cancer comes directly following the birth and during the breast feeding of her twins makes this relationship even more uncanny. In his "After Words," her husband, Matt Seaton, using almost the same language as Rosenblum uses above, recognizes the significance of these proximal—both in time and in sensibility—embodied events for Picardie:

> Perhaps this is fanciful hindsight, but I sometimes feel that the arduous experience of her twin pregnancy taught her, if almost imperceptibly, to mistrust her body, to feel that it was not the constant, easy friend it had been. She used to say herself that she felt the foetuses she was carrying were somehow inimical to her—as if they were parasitical aliens like the Midwich Cuckoos. It had nothing to do with how she felt about having children; about that she was unambivalently positive. It was simply how she chose to describe the physical sensation of being pregnant. (1998, 102)

In her *Observer* column, Picardie rather sheepishly admits to her readers that, since her cancer diagnosis, she has become convinced that she is in fact pregnant again. Denying that such a lie of the body is simply a peculiar—and not very evolved—form of denial, Picardie explains to her readers the rationality of such an uncanny association: "I need only to refer you to one of the pregnancy manuals dusting up my shelves: the vomiting, the weird stuff growing inside you, the endless waiting for the big day" (84–85). In a sense, pregnancy prepares her for the experience of cancer, though, of course, the big day promises death not life when it finally arrives.

Cancer in Two Voices and *Before I Say Goodbye* take up the transformation of the body generally in illness, and they also attempt to describe, in particular, the transformation of sexuality that takes place; or, more precisely, they describe the *desexualization* that often occurs along with illness. Seaton, recalling the "delicious feel" of Ruth's breast in his hand, is conscious of the irony of eros and thanatos "gorgeously cupped in my palm" (102). He admits that after the diagnosis he found it "unbearable to touch her there," and acknowledges that "the lump not only grew within her, but between us, spreading, as inexorably as the cancer itself did" (103). Picardie herself explains this process of desexualization in an email to her friend India: "My libido is DEAD. Most of my pubic hair has fallen out because of the chemotherapy. I look like an alien. Yuk yuk yuk" (44; emphasis in original). Desexualization occurs in a slightly different form between Butler and Rosenblum. As Rosenblum puts on weight during her treatment, Butler is "unthinkingly delighted as her hips spread and grew wider" (140). Butler attempts to convince her partner that her body is still sexy, that, in fact, it is—in its new, fuller form—even more sexy than before. Rosenblum, however, is unable to experience her new body in terms of desire, and, thus, Butler's response is somewhat jarring and feels forced, an example of how she wants to feel, or thinks she should feel, for Rosenblum's sake.

Although the cause of both Rosenblum's and Picardie's desexualization is not an intentional redirection of libido into creative work, but an unintentional by-product of the physiological effects of cancer treatment, nonetheless, to some extent, the desexualization that occurs in illness allows—or requires—a concomitant sublimation of sexuality into creative expression. Leo Bersani has called sublimation an "eroticizing of consciousness in art" (1986, 43), which is similar, I think, to Lorde's conception of the "erotic," and her opposition, discussed in *The Cancer Journals*, to what she calls "false spirituality." "False spirituality," according to Lorde, "is false because too cheaply bought and little understood, but most of all because it does not lend, but rather saps, that energy we need to do our work" (1980, 39). For Lorde, transformation requires work: political, creative, and otherwise; it requires making use of the power of the erotic, which is concerned with what we do and with "how acutely and fully we can feel in the doing" (1993 [1978], 340). Lorde's erotic, which also connects to Gillian Rose's notion of "love's work," which I will discuss in my conclusion, links doing and feeling, the political and spiritual, the effective and affective. Sublimation, then, as Freud conceived it, is as much an ethical as an aesthetic act. As Bersani shows with regards to Freud's own work, sublimation is not, or not only, a phenomenon in which desire is denied; it is about making something out of one's desire, not simply a giving up of one's desire. Desire, as we shall see in this chapter and the next, can be diversified rather than denied, and illness has the capacity to unbind the ego from the internal super-ego, and from the external, cultural super-ego as well. The experience of illness can lead to an inventiveness that is not necessarily about the formation of a fixed identity, but about queering identity and inventing new forms of meaning, being, and doing, as Sedgwick's work calls for.

Another aspect that connects Rosenblum's experience of cancer with Picardie's is the fact that both cases seem especially tragic because, apparently, neither case had to end so quickly in death. Both cases, in fact, reveal conventional medicine as particularly fallible. Rosenblum was misdiagnosed and mistreated from the beginning, and her life would certainly have been extended if not for medical incompetence (1991, 11). *Cancer in Two Voices* documents not only her illness but also her bringing of a malpractice suit against Kaiser Hospital. Kaiser settled her case not long before her death for $296,000, plus $25,000 a year, increasing 3 percent annually for the length of Rosenblum's life. As she notes, this reward is "blood money," and, although it makes her "very rich" financially, it doesn't dissipate her "rage and tremendous sadness" (18). Picardie's family also received some financial compensation—a £300,000 out-of-court settlement from Guy's and St. Thomas's National Health Service Trust, whose doctors, like Rosenblum's, diagnosed her disease too late. According to the *Guardian* health correspondent, Sarah Boseley, the claim was settled in a "comparatively speedy two years in tacit acknowledgement that doctors were wrong to send Ms. Picardie home with reassurances that a breast lump was benign in 1994 after tests which were not conclusive" (Boseley 1999, 6).[3] Lorde's materialist assertion that I cited in the last chapter is prophetic where Rosenblum and Picardie are concerned: "We live in a profit economy and there is no profit in the prevention of cancer; there is only profit in the treatment of cancer" (1980, 71).

Money, as Rosenblum's and Picardie's families well know, is not compensation for Rosenblum's death at 44 and Picardie's death at 34. And so, *Cancer in Two Voices* and *Before I Say Goodbye*, like Lorde's *Cancer Journals*, attempt to offer a sort of narrative compensation. In doing so, both breast cancer journals try as well to break free of more conventional autobiographical narrative forms by utilizing the journal form to record day-to-day observations of the physical and emotional crises that characterize a life-threatening illness, and incorporating other voices into their narratives. *Cancer in Two Voices* creates a dialogue between the person with cancer and her lover and primary caretaker, and *Before I Say Goodbye* mixes email correspondence[4] between Picardie and friends, Picardie's weekly newspaper columns about her experiences of illness, letters from readers in response to her columns, reproductions of Picardie's handwritten notes to her children, a newspaper column completed by her sister after her death, and further "After Words" from her husband, Matt Seaton. These polyvocal representations of the experience of breast cancer both succeed and fail. Although they succeed in communicating the social aspects of illness, they nonetheless fail in fully communicating the existential condition of terminal illness,[5] because, ultimately, it is a condition that cannot be shared or communicated in any positive way. And yet, despite this failure—this "crisis in witnessing," as Felman and Laub (1992) have called it with regards to Holocaust testimony—revealed by both of these breast cancer journals, the sense that "words matter," as Seaton says, and that "silence must be transformed into language and action," as Lorde says, permeates them both. How might the body in crisis be brought to language and communicated to others, and what are the ethical implications of doing so? How might both loss (of a breast, of a loved one) and the mourning of that loss be articulated?

As I noted in the last chapter, Sontag, Lorde, and Sedgwick provide counternarratives to the clinical medical narratives of the experience of illness. I want to reiterate here that there is not one, single, hegemonic counternarrative, but many. In the rest of this chapter, again following Foucault's methodology, I describe and juxtapose two different arts of being ill, which I see as emerging out of different cultural milieu and with different publics in mind. While it is clear that illness narratives have enjoyed a measure of popularity in both the United States and Britain in the past twenty-five years, it also appears that the types of illness narratives published and the reasons for their favorable reception are quite different in each country. I am interested, therefore, in the possibility that narratives of illness and death and their reception reflect contrasting national attitudes and/or ideologies about illness and death, and that they contribute to the creation of imagined national communities of the healthy and the ill. The arts of being ill that I identify for the United States and Britain do not by any means apply to each and every illness narrative published and read in these countries throughout history. Rather, I suggest them as fictions that structure, to a greater or lesser extent, conceptions of the self in these countries at a particular historical moment. Indeed, one might read Sontag's critique of metaphorization of illness as also concerned with national arts of being ill. She called these metaphors "sentimental fantasies," and suggested they represent "not real geography, but stereotypes of national character" (1978, 3). Here I want to diagnose two of these "sentimental fantasies" and "stereotypes of national character," and the ways that they come into being in relation to each other. How does the perception of a national character or self—and, concurrently, of a national Other (an-Other nationality)—help situate a nation's own narratives of illness and death? How might such differences be expressed in terms of content and narrative form? In attempting to work out the nature of such differences it is important to bear in mind what Elizabeth Spelman has called a "rhetorics of suffering"—that is, whose suffering counts (and is published and read) and whose does not—that is always at work in these narratives.

The Cultivation of an Improved Self

In her essay in the *New York Times Book Review* on the Oprahfication of literature in America, Martha Bayles writes about what might be seen as a resurgence of reading in the video age. Bayles claims that the difference between people who frequent Blockbuster Video and Barnes and Noble in places like Metairie, Louisiana, "is not some McLuhanesque gap between print and electronic culture but rather the difference between pandering to the lowest common denominator and offering people something uplifting" (1999, 35). Bayles continues:

> I use the word "uplifting" advisedly, knowing that it carries a load of condescension. But that is precisely my point. The secret of Winfrey's golden touch is her willingness to shoulder the mantle of cultural authority—which, ironically, most literary people seem desperate to shrug off. She can do this because her appeal to viewers has never been based on the exploitation of their social and emotional problems (in the manner of Jerry Springer) but on self-help: a trendy term that nonetheless subsumes older and more deeply ingrained notions of self-improvement and (yes) moral uplift.
>
> Winfrey's [book club] selections reflect a coherent sensibility: she favors novels about people, mostly women but some men, who are up against hard circumstances but who manage to endure, if not prevail, in the spirit of what Ralph Ellison called heroic optimism. (35)

Bayles is describing the cultivation of a particular kind of self through reading: a self concerned with self-improvement and moral uplift.

The sensibility that Bayles asserts as characteristic of Winfrey's book club selections can also be found in many American memoirs. In an analysis of the "Oprah Effect" for the *New York Times Magazine* in 1999, D. T. Max considers the "therapeutic canon" that Winfrey's book club has established. As Max understands it, the "real innovation that allowed Winfrey to turn novels into TV" is that she and a group of viewers discuss each book in terms of its "relevance to their own lives" (1999, 36). Max finds "Winfrey's insistence on treating novels as springboards for self-revelation" odd, because it implies that reading is about delving deeper into oneself rather than about stepping outside of oneself to read about the experiences of others. For Winfrey, reading is a therapeutic practice of the self in which the literary other is identified with and incorporated into the self. In a sense, then, reading, in Winfrey's conception, is not about the self in relation to others, but rather about the self in relation to itself. Max quotes Winfrey as saying to her audience, "The reason I love books . . . is because they teach us something about ourselves" (37) and "Reading is like everything else. You're drawn to people who are like yourself" (39).

In an earlier *New York Times* article on the Oprah phenomenon entitled "Inner Peace, Empowerment and Host Worship," Jeff MacGregor notes that in its 1998 incarnation the "Oprah" show "has, like its host, undergone yet another makeover" (1998, 30). According to MacGregor, the makeover has resulted in nothing less than a "psychospiritual Reformation." This makeover, it seems to me, extends the therapeutic practice of the self that Winfrey began in 1996 with her book club. In Winfrey's "change your life TV," the protagonist to be drawn to as someone who might teach us something about ourselves is none other than Winfrey herself. As always, Winfrey is, as MacGregor is well aware, quite savvy in ascertaining as well as, paradoxically, helping to define a late-twentieth- and early-twenty-first-century American art of existence:

> The show is devoted solely to the mind-numbing clichés of personal improvement. It is a self-help anthology, presented with the missionary zeal of someone who's already got hers. Thus it is about nothing so much as Ms. Winfrey herself, and her pilgrimage toward a more rewarding state of Oprahness. Given our nation's endless search for the short cut to our better selves, it is a brilliant programming move. (30)

MacGregor sees Winfrey as a contemporary guru and a latter-day circuit rider, who has "found a way to shamelessly market the history of her own misery and confusion as a form of worship" (30). He acknowledges that her "solipsistic approach to self-betterment borrows literally from the world's great religions," and what worries him isn't Winfrey's guru status per se, but, rather, that "this kind of navel-gazing seems to result in a myopic national selfishness wherein we trade our commonly agreed upon moral imperatives for the short-term comforts of our new incontestable individual emotions. (I feel, therefore I am.) We want to feel good about ourselves, or at least better, sometimes at the cost of our best collective interests" (30).

My digression into the Oprahfication of literature is intended to begin to describe an American art of existence, or style of imagining, which emerges not only in the person of a contemporary cultural figure like Winfrey, but also in national myths that create an imagined "American" community. In his essay on the origin and spread of nationalism, Benedict Anderson defines a nation as "an imagined political community," and asserts that such "[c]ommunities are to be distinguished, not by their falsity/genuineness, but by the style in which they are imagined" (1991, 6). Anderson links the emergence of the nation in the eighteenth century with "two forms of imagining" that also developed at that time: the novel and the newspaper (24–25). The novel and newspaper, according to Anderson, "provided the technical means for 're-presenting' the *kind* of imagined community that is the nation" (25; emphasis in original). Using Anderson's reading of the connection between forms of imagining and the emergence of "imagined political communities," I wonder what illness narratives might have to do with the nation and its imaginings.

Anderson is interested in understanding various forms of imagining—not only the novel and the newspaper, but also autobiography, which he touches on briefly in the revised 1991 edition of *Imagined Communities*—as "cultural products." Anderson notes that the practice of reading the morning newspaper is a mass ceremony that allows the reader to be "reassured that the imagined world is visibly rooted in everyday life" (35–36). The same sort of mass ceremony, it seems to me, is performed by Winfrey's reading group, as well as by the mass consumption of memoirs, set, like the novel and the newspaper, in homogenous, empty time, which allows a metahistorical and metageographical connection between readers (204). Memoir, at least as it is imagined, is a radically democratic literary style; it is something of a vernacular form of representation. Illness narratives might, therefore, be thought of as the patient's vernacular, signaling the vernacularization of science and medicine and its sacred language.

As Bayles's reference to Ralph Ellison makes clear, an American mode that emphasizes self-improvement and moral uplift is by no means a recent phenomenon. Winfrey herself might be considered in relation to a Protestant religious tradition as well as national myths that promote capitalist values as moral values. In his cultural history of psychotherapy in America entitled *Constructing the Self, Constructing America*, Philip Cushman locates the emergence of a philosophy of "healing through self-liberation" in the mid to late nineteenth century, as exemplified by the mesmerism movement, and the "mind cure" movement that followed mesmerism. According to Cushman, mesmerism was characterized by four theoretical concepts: apolitical interiority, cognition, the true self–false self dichotomy, and the idea that mind cure led to, or was made manifest by, material abundance (1995, 128–29). Although Cushman recognizes that the "configuration of self in the late twentieth century is somewhat different from that of the nineteenth century," nonetheless, he believes, and I would agree, that the overall schema he identifies structures the current American configuration of self: "the emphasis on interior, insular cognitive processes, the focus on the well-being of the self, and the foundational belief in the liberation of the enchanted interior" (129). Cushman diagnoses the shift from a configuration of the self concerned primarily with spiritual health to one concerned primarily with mental health and material wealth. In the late twentieth century, we see another shift in the American configuration of self, this time to a configuration of self that is concerned primarily with bodily health as an outward manifestation of spiritual and moral health.[6] It is the cultivation of precisely this kind of self that we see at work in *Cancer in Two Voices* and many other American illness narratives.[7] These draw on an American autobiographical tradition concerned with education, both moral and intellectual, and progress, of the self and of the nation, as exemplified in two classic American autobiographies, Benjamin Franklin's *Autobiography*, originally published posthumously in 1791,[8] and Henry Adams's *Education of Henry Adams*, originally published in 1906.[9] While I take this particular configuration of self to be hegemonic in American culture, as my conclusion suggests, counternarratives do emerge that challenge this dominant American art of existence.

Cancer in Two Voices opens with Barbara Rosenblum's death and Sandra Butler's loss; Butler writes, "On Valentine's Day, February 14, 1988, Barbara Rosenblum died. She was my partner, my friend, my love." Quickly, however, Butler moves from death and loss, with a call to readers to enter Butler and Rosenblum's world, to learn to "love more fully and deeply," and, presumably, to be "uplifted" in the process, because theirs is a story, according to Butler, "about struggle and courage, even more than . . . about sickness and death. A story about loss and the gifts it brings" (1991, i). Butler explains that when Rosenblum told friends in February 1985 about her diagnosis of advanced breast cancer, she understood that "she was only the first" (i). Rosenblum perceives in the moment of such a grave diagnosis that many of her "friends will see their future in the way I handle mine" (12). Thus, she and Butler emphasize from the start the need to live self-consciously and in an exemplary manner. Cancer becomes, quite simply, an opportunity—a gift, even—for them to live better. According to the heroically optimistic Butler: "[T]his isn't just a book about cancer, and it isn't really a book about dying either. It's a book about living expansively, openheartedly, joyfully in the face of our inevitable death. Instead of looking for miracles we can each become our own miracle" (iii). And so, like Lorde, who understands illness as a "corporate effort" (1980, 30), Butler and Rosenblum, in the face of their uncertain future, "mobilize a battalion of friends" to assist Rosenblum in "every phase of fighting [her] disease" (11–12). Illness, then, is an opportunity to create a certain model of relationship—a "coupling of existences," as Foucault would say—which enacts a "mutual solicitude" (1986a, 152), not only between Butler and Rosenblum, but beyond them (including, of course, the readers of their book).

For Butler and Rosenblum, then, cancer offers a challenge, to which they respond with earnestness and creativity. "I have cancer but it is not consuming me," Rosenblum assures her readers, and herself as well. "Rather, I am as alive as I can be; my creative juices have never been as electric; my thoughts have never been as clear. With each new status in my illness, my range of choices shrinks but I become deeper and richer, clearer and simpler to myself" (56). The simplification that illness brings for Rosenblum is, as she understands it, a coming to one's essence, a coming to one's "fullest and truest self"; it is "[t]he self revealed to the self, observed by the self" (105). For Rosenblum, through illness, the self that is revealed to and observed by the self must also be told. The story Rosenblum tells, then, is a *story told for the self,* and with a "desperate sense of urgency" (122).

As Rosenblum nears death, she and Butler attempt to maintain control over their chaotic and unraveling lives. The story told for the self is an exercise in control, a noble attempt not to be consumed by the disease. And yet, their control over the illness and over their story of the illness falters; what seeps into the narrative is both a repressed and a not-so-repressed anger. Despite efforts to make it otherwise, cancer becomes what both Butler and Rosenblum do, and who they are. They are stymied in their attempts to invent new forms of being and methods of doing that take them outside of and beyond Rosenblum's cancer. Butler admits bitterly, "Cancer swallows up the air of my life and insinuates its presence everywhere. Nothing remains untouched" (48). And Rosenblum despairs:

> Massage. Acupuncture. Herbs. Relaxation. What more can I do? My life feels worthless. I feel depressed and I can understand wanting to die. H., in her work with people with AIDS, says many die angry. I'm trying not to. I'm trying to achieve completion. Maybe the strain is too much. (79)

One might ask of Rosenblum, which strain is too much?: that caused by her illness and the anger that accompanies it, or that caused by trying not to be angry and "to achieve completion"? What happens when the better self one seeks to cultivate simply doesn't get better? As Ehrenreich, Lorde, Butler, and Rosenblum all attest, that self often gets angry. Unlike Lorde, however, who is compelled by the memory of those who did not survive to always direct her anger outwards into political struggle, Butler and Rosenblum often helplessly direct their anger inwards, at themselves and at each other. Mutual solicitude threatens to become mutual decrepitude as the possibility of redemption is overwhelmed by the reality of dying. It isn't that Butler and Rosenblum submit to what I described in the last chapter as the discourse of responsibility; they know who is to blame for Rosenblum's misdiagnosis and her death that comes too soon. Rather, in their need to focus on the good that cancer brings, the possibility to improve one's life and redeem oneself, they seem almost desperate to screen the real with a story of heroism—a story for the self—even as the self that this story is for disappears from the scene they have constructed.

The Cultivation of an Ironic Self

Butler and Rosenblum's rather earnest and somewhat humorless tone (cancer, they seem to say, is a deadly serious business) is in sharp contrast to Picardie's ironic and self-deprecating tone (there is a danger, she seems to say, in taking oneself too seriously, especially when one is terminally ill). In an interview with English playwright and novelist Michael Frayn, Matt Seaton, Picardie's widower, describes Frayn's work as that which "carries English self-deprecation to the level of self-flagellatory performance art." According to Seaton, Frayn's talent lies in his ability in telling "the story told against oneself" (1999, 8). This is precisely what Picardie tells in *Before I Say Goodbye*, a *story told against oneself*, and this is what makes her work so different from Butler and Rosenblum's and other recent American illness narratives,[10] where one rarely finds stories told against the self because it would be seen as adding insult to injury. In many contemporary British illness narratives,[11] however, the story told against oneself is used to reveal vulnerability and fear through humor that hides, but never very well, the "threat of something quite painful" (8).[12] Unlike the story told for the self, the story told against the self does not create an economy in which loss must be converted into gain; rather, it holds onto loss, complicating it with humor, perhaps, but not converting it, or simplifying it into its opposite. In a cover story in the *New York Times Magazine* entitled "There Will Always Be an England," Andrew Sullivan argues that, "Loss . . . is the central theme of modern Britain: loss of empire, loss of power, loss of grandeur, loss of the comfort of the past" (1999, 78). I think one might push Sullivan a bit by asking if everyone in modern Britain experiences this loss (of empire, for example) in the same way, and, moreover, whose grandeur or whose comfort has been lost. (The same question, of course, could be addressed to my own work, and to some extent will be considered in the following two chapters.) But, nonetheless, Sullivan's description of a contemporary British "culture of loss" connects with what I take to be a modern British mode of being ill as exemplified in published illness narratives.[13] Cushman, like Sullivan, also notes that the European proximity to loss beginning with the First World War, and the American geographical and emotional distance to such loss, has meant that "the collapse of the moral underpinnings of the modern era did not appear to be felt as strongly in the United States" (1995, 164).

For Picardie, then, cancer becomes not an opportunity to live better, but an opportunity to resist the sentimentalization of suffering. Lorde, too, resists sentimentalization when she explains, "Pain does not mellow you, nor does it ennoble, in my experience" (1980, 49). But, contrast Lorde's unsentimentality to Picardie's. She acknowledges that "diagnosis turns you into a grumpy, bitter, envious old cow" (38), and notes, in an email to her friend Jamie, two "fun things about breast cancer":

> 1. You get your hair cut really short because it's falling out, and it really suits you. You decide to keep it that way forever.
> 2. You can be really horrible to people and not feel guilty. (1998, 4–5)

It is not that Picardie's illness is not transformative; the changes in her body and her inexorable movement toward death at a young age are clearly a crisis of subjectification for Picardie and Seaton as much as they are for Butler and Rosenblum. Nonetheless, she exhorts Jamie, who is HIV+, to "[r]esist the temptation to come over all American and blurt out 'I love you' to all your friends" (16). Picardie's British art of being ill conceives itself in relation to an American art of being ill; the American art of being ill is viewed with both wonder and cynicism, but is to be avoided at all costs.

When I have presented my work on British and American arts of being ill and doing illness in both the United States and the United Kingdom, I have had on occasion divergent responses. Interestingly, responses have tended to follow national lines: while most Brits have been enthusiastic, some Americans have remained unpersuaded. I have also noticed that cultural studies scholars have been fairly enthusiastic over the comparison because it concerns the cultural work that literature does, whereas literary scholars think my claims are too substantial based on a case study of two texts. I submit that in contemporary Britain there is a widespread belief that there is a particular mode of self-writing that is "American," and when British writers try to write against it or outside of it, they are often consciously challenging current American hegemony in popular culture. Americans are rightfully somewhat dismayed to have their literary productions reduced so drastically, because of course not all American writers, nor even all American memoirists, write in an earnest and heroic mode (no more than all British writers write in an ironic mode). However, when Ruth Picardie writes that she wants to avoid "coming over all American," her readers know what this looks like. By asserting that there is an American mode of being ill and doing illness, I do not mean to imply that this mode has remained stable throughout history. Rather, I believe this mode has emerged at a particular time in which American culture is perceived by many people inside and outside the United States as hegemonic, and, since America's "winning" of the Cold War, as heroic. The American mode of being ill demonstrated in many, though not all, illness narratives continues an American autobiographical tradition, but it also represents America's position in the world today, and this contrasts to Britain's position in the world today (and Britain's contemporary position contrasts with its former imperial position).

Recently, there has been something of a backlash against confessional culture in Britain, and many British writers and critics have distanced themselves from the confessional mode by portraying it as an American import. For example, an article in the *Daily Telegraph* about Martin Amis's memoir *Experience* (Johnson 2000) associates Amis's (misguided) decision to write a memoir with his (equally misguided) preference for America over Britain. Amis, we learn, "is reported to be threatening, yet again, to move to America." Thus, Daniel Johnson exclaims, "Martin: you have disappointed us. We had intended you to be the next Nobel laureate but three. The rights were sold; the press was squared; the middle class was quite prepared. But as it is . . . the notion grates. Go: Kingsley's Garrick chair awaits!" (20). Go where? To America, of course. To do what? To write memoirs, of course, because that's what they (Americans and, perhaps, would-be Americans like Amis) do there. Further establishing the contrast between national styles of literary imagining, Johnson also complains that Americans can't appreciate the British novel, an observation that serves to reinforce his view that the memoir is a (degenerate?) American form of imagining, while the novel is a (virile?) British form of imagining. He ends with the assertion that, "[i]t isn't where you live that matters; it's what you write" (20), again ironically implying that memoir is an American genre no matter who writes it.

Another British review of Amis's *Experience* by D. J. Taylor in the *Independent* also implies a direct connection between memoir writing and America, attributing the British disinclination to confession to lingering Victorian reticence: "The writer's confessional is a fairly recent development—at any rate on this side of the Atlantic. Many a Victorian novelist, and many a Victorian novelist's public, would have been horrified by the thought of even a respectful posthumous biography" (2000, 5). Taylor, nostalgic for this supposed Victorian reticence in the age of memoir, is certain that "this cavalcade of personal revelation . . . is likely to have deeply depressing consequences for the art, or rather *the way in which the art is received*" (5; emphasis added). Taylor is troubled by the meshing of art and experience in the age of memoir, which is not unlike Arlene Croce's argument against what she calls "victim art" that I discussed in my introduction.

The backlash in Britain against confessional culture in general, and illness narratives in particular, is also highlighted by the fact that, from May through June 1999, the *Observer,* the paper that published Picardie's columns a year earlier, published a series of spoof columns entitled "Time to Go." The first piece opened with a note from the editor stating,

> Last week, *The Observer* learnt that columnist Richard Geefe had attempted to take his own life. Despite his need for rest and medication, he insists that he continue writing and I have, after much thought, agreed. We will, of course, review the situation continually in line with Richard's true interests. For *The Observer* prurience is and always will be inexcusable. We have also agreed not to alter his work in any way, however uncomfortable that makes us feel. (Geefe 1999a, 2)

Geefe's column was, according to *Observer* editor, Roger Alton, "a Swiftian satire on the confessional column." It was also clearly a send-up of Picardie's *Observer* column, as well as of John Diamond's columns in the *Times* about his throat cancer (Minnie 1999, 21). For example, in a spoof of one of Picardie's "fun things about cancer," Geefe (who is really the British TV and radio hoaxer Chris Morris) relates a scenario in which he and a "four-year-old git called Ivan" vie for the last smoked cod's roe cracker at a picnic arranged by friends to cheer Geefe up. Geefe behaves badly—in fact, far worse than his four-year-old rival—and selfishly wolfs down the cracker in one gulp rather than share it with Ivan as has been negotiated by some of the other picnickers. Geefe feels no shame, realizes he has "been excused because of [his] 'condition,'" and wonders "[w]hat the hell else [he] will be allowed to get away with" (Geefe 1999b, 4).

Geefe's column is particularly cutting with regards to those readers who are moved by other people's stories of illness, calling such readers "sobbing empathy boffs" (1999c, 4) and "weirdos who've set up a daily Geefe vigil in the pub on the corner" (1999d, 4). He notes acerbically, "When 40 independent strangers tell you how brave you are, it is difficult not to suspect they are right" (1999c, 4). Picardie too satirizes those people who see her as their "favorite cripple," and who feel they are "entitled to regular blow by blow accounts of [her] emotional/physiological state" (1998, 18). In her column, she calls people who find terminal illness exciting "road accident rubber-neckers" (73). But, Picardie is clear that such rubber-neckers are not necessarily people who might read her story and be moved by it, but rather people who she has met in the past who suddenly come out of the woodwork in order to become "secular Samaritans looking for glory" through their relationship with someone who is dying. This may seem a fine distinction between Geefe's "sobbing empathy boffs" and Picardie's "rubber-neckers," but it seems clear that Geefe is satirizing those who claim to feel something upon reading about another person's crisis of the self, whereas Picardie is satirizing those who think they might benefit from a relationship with that person. Geefe is suspicious of the possibility of empathy through reading, while Picardie is suspicious of those who attempt to take advantage of their relationship with her to feel better about themselves.

The writers and readers of confessional columns are not the only ones subject to Geefe's satire, however; he also sets his sights on those publishers (newspaper and book publishers, as well as broadcast media) who have benefited financially from the confessional craze, and have, thus, in a sense, created the genre itself:

> The moment I croak, these columns will be on sale for 16.99 a pop to a lot of people who've already read them. But in order to catch the Christmas rush, the publisher needs all copy by the end of August. So I am having to write the end of my life now committing myself to what I'll be doing, how I will feel about it, and my exact method of blapping my lulu. It's driving me nuts. (1999d, 4)

This, too, is apparently a direct reference to Picardie, who mentions in an email to Jamie, "I definitely want our correspondence to be published posthumously. (Lots of money to our widowers.) How about The Diary of Ruth and Jamie. 'In 1997, two young people [I'm going to be 29 in my obits] went into hiding in South London when their bodies were occupied by invading Bad Cells. This is their moving diary... Blah blah blah'" (16). Even before Geefe's "Swiftian satire on the confessional column," it seems, Picardie herself was already spoofing the genre.

Geefe's column sparked a flurry of responses in the British press, including one from some *Observer* staff members who called it "extremely misjudged" (Minnie 1999, 21), as well as a rather, dare I say, *earnest* discussion of the potential value of confessional literature. In her June 29, 1999, column in the *Guardian*, Linda Grant entered the fray as both a journalist and someone who has written her own memoir about her mother's dementia and the ethical dilemma she faced over whether or not to put her mother into a nursing home. Grant acknowledges the fact that much confessional literature does "trivialise and sentimentalise private life," and, furthermore, that it does "encourage voyeurism and the construction of vicarious relationships with people the readers will never meet" (1999, 13). But Grant discovers in the process of writing her own memoir and in its public reception another function that confessional literature performs. According to Grant, simply put, such work can help people. Thus, she castigates other critics for their cynicism, noting that,

> [r]eaders of these books and columns are being told that what they feel when they read them is somehow bogus because the critics don't feel the same way. You can have any number of arguments with people about what they think, but when you start telling them what to feel, then you're really walking on thin ice. (13)

Grant puts her finger on the gap between thinking and feeling that confessional literature attempts to bridge or collapse, whereas the critical response to such literature often attempts to maintain this gap, aligning thought on the side of critique and feeling on the side of confession, a distinction that Sontag seemed to also want to maintain in *Illness as Metaphor*.

Geefe's column and the plethora of responses to it reveal that in Britain, unlike in the United States, much confessional writing is being published in the national newspapers by already established journalists.[14] In a sense, then, in the British context, confession has become a journalistic phenomenon as much as, or more so than, a book publishing phenomenon. What becomes problematic then with regards to Geefe's column is not its send-up of confessional literature per se, but rather a feeling one gets when reading it that its intended audience is not the public who might read these columns ("those sobbing empathy boffs") at all, but *exclusively* other journalists, creating a particularly insular form of writing, whose purpose is to reveal how oh-so-clever "we journalists" can be (wink, wink, nudge, nudge). Thus, in an article in the *Times* of London defending Geefe's column, Stephen Armstrong opens with a comment only someone who is, or would like to be, in the in crowd of clever journalists could write: "On reflection, it should have been obvious it was Chris Morris. The idea of a journalist announcing he was going to commit suicide in six months' time but would continue to write a column every week until he finally popped his clogs . . . pure Chris Morris. The language. That name. Those turns of phrase. The mutant words—'chonk,' 'morbunderast,' 'twunt'" (1999, 10).

The *Observer* spoof is interesting, in particular, if we juxtapose it to the case of an American reporter, Kim Stacy, who fabricated having brain cancer and wrote a series of columns for the Owensboro (Kentucky) *Messenger-Inquirer* about her "experience" of illness. When it was discovered that Stacy didn't have brain cancer, she said that, in fact, she had AIDS, but had lied about the "real" diagnosis because of the stigma surrounding AIDS. In the end, her colleagues at the *Messenger-Inquirer* and her readers learned that she wasn't ill at all, unless one lets stand psychologist Patty Randolph's diagnosis of "factitious disorder," a disorder in which people lie about their health in order to get attention (Poynter 1999, 2B).[15] Stacy created her own illness not to spoof the craze for confession, but to buy into it in all seriousness, adding fuel to the many British critics (and some American critics as well, including Croce) who see confessional literature as, at best, trivial and sentimental and, at worst, simply a big lie masquerading as revealed truth. Michael Ellison's report in the *Guardian* on Stacy's lies tellingly focuses not on Stacy's alleged betrayal of the truth itself, but on the fact that Stacy "embraced the cult of the confessional columnist with obsessive authority" (1999, 15). In the British context, cleverness is prized over earnestness, which is perceived, as Picardie herself understood, as "coming over all American."

Picardie's resistance to coming over all American, however, doesn't prevent her from asserting a politics of breast cancer based on what she calls the "HIV/AIDS paradigm" (1998, 15), which emerged as I showed in chapter 2 out of the American context.[16] In an email to her friend India, she admits that her diagnosis of advanced breast cancer "is a total headfuck." But, she continues,

> I think the HIV/AIDS approach is a good one ie accept that I have a terminal illness, that I am not going to live for three score years and ten, and then be positive about maximising length and quality of time I have left. That means keeping healthy, embracing orthodox and complementary treatment, and having a good time.[17] Cancer is all about fear, secrecy and euphemism—palliative care, advanced disease—all are euphemisms for dying. Oncology is the biggest euphemism in the world. But there is another way. (13)

Opposing euphemistic speech with true speech, Picardie's politicization takes the form of straight-talk about her illness in her weekly column for the *Observer*, where she doesn't cultivate an improved self (though she'd damn well like to get better), but an ironic self:[18]

> And that lump in your left breast, the one you noticed after you stopped breastfeeding last summer? In the . . . Clooney-filled scheme of things, the hospital would smile and tell you not to worry, it was the harmless fibroadenoma they'd found in 1994. But this is the fat, stained, piggy-eyed parallel world of illness, and your lump, I'm sorry to say, is actually cancer. Or should we say lumps, because, oops, it's spread to the lymph nodes under your arm and in your neck, which means it's stage three cancer and you've a 50:50 chance of living five years. (38)

Picardie refuses to offer the consolation of an *ER*–style redemption to her readers; and, even more importantly, she refuses to seek such consolation for herself. In his "After Words," Seaton continues Picardie's quest for true speech, admitting that "the illusions one holds about a peaceful, dignified death and the family's perfect bedside farewell will almost certainly be tugged away" (100). Seaton's lesson before Ruth's dying is that "[d]ying is nasty, ugly, and painful; it's so obvious, isn't it?" (100). The question after the statement reveals his unwillingness to accept that this is all death is, and it reminds us, paradoxically, that there isn't anything obvious about death at all.

What is consoling for Picardie, however, is the possibility—and the therapeutic effects—of a "new life as a bimbo" (28), in which her practices of self improvement are, contra Lorde and Rosenblum, quite literally, cosmetic. In her *Observer* column, Picardie recommends "retail therapy"—that is, "personal indulgence or escapism of any kind"—to her readers, noting that it is cheaper than alternative medicine, more fun than chemotherapy, and actually effective (78). Spoofing the discourse of personal responsibility and the neoliberalization of the experience and event of illness, Picardie notes that she intends to write a book, *Shop Yourself Out of Cancer,* to counter the New-Age books weighing down her shelves: "*Sharks Don't Get Cancer; Living, Loving and Healing* (written by wise gnome with magic hands); *Full Catastrophe Living* (inspirational picture of high mountain); *Spontaneous Healing* (written by man with beard)" (77).

While Rosenblum experiences breast cancer as "a step towards clarity, rather than limitation," Picardie, too, knows that "terminal illness is supposed to make you extremely wise and evolved" (48). Yet, she is suspicious of "all this interest in sick people," and admits "wanting to stay as far away from them as possible" (72). She acknowledges that she can't quite get her "head around Zen meditation," and confesses, rather, to be stuck in a "Why did I eat the fish fingers that Lola spat out when I can't fit into my jeans anymore?" mode (48). Nonetheless, her illness and her writing about it does make Picardie able to reach out to others, if only because, as a reader of her column writes in a letter to the *Observer*, she "reminds us that this is happening to a real flesh-and-blood human being, and not one who has been mysteriously transformed into some serene apprentice angel with a foot in two worlds" (55–56). It may not be Picardie's style, her art of being ill, to say "I love you" to all her friends, but her writing creates a "coupling of existences," a "mutual solicitude" with her readers, and this is reflected in the text of *Before I Say Goodbye*, which incorporates numerous letters from readers, who send Picardie their stories, their fears, their prayers, and, yes, their love. Picardie, therefore, offers an "ironic passion" that is unlike Butler and Rosenblum's earnest passion or, for that matter, Geefe's ironic dispassion.[19]

The Art of Dying and the Failure of Stories

I conclude this chapter by considering illness not in relation, but in isolation. I move, therefore, from the plural arts of being ill and doing illness to a singular art of dying. This is not to say that everyone dies in the same way; it is only to say that everyone dies alone. There is in illness and dying, according to Lorde, a "commonality of isolation" (1980, 10). On the one hand, illness narratives demonstrate the cultivation of a particular kind of self, while on the other hand, in the case of narratives of terminal illness in particular, they demonstrate the unmaking of the self. Illness narratives present the cultivation of self and, finally, its failure. In these narratives we readers are confronted with many failures: the failure of the body, the failure of conventional and alternative medicine, and, in the end, the failure of stories in the face of death. Rosenblum recognizes, for example, the impossibility of her narrative task: "What can I write about this aloneness?" she asks, knowing there is nothing to write about it, nothing in the experience of living with which to compare it (1991, 119). And Seaton recognizes that "[c]ancer changed everything" between him and Picardie: "it put us on different tracks, stretching our grasp of one another to the limit and essentially forcing us apart. In the end, I could not reach her, and it felt like a failure to me. And then she was gone" (103–4).

Before taking up the rendering of death in *Cancer in Two Voices* and *Before I Say Goodbye*, I want to return, briefly, to *The Cancer Journals*. Although, as I noted in the last chapter, *The Cancer Journals* encourages the integration of death into living, it does not contain Lorde's actual death in its original version.[20] However, a special edition of *The Cancer Journals* was published in 1997 that ends with twelve tributes to Lorde that had been published previously in *Sojourner* just after Lorde's death in 1992, as well as six photographs taken by Jean Weisinger at the "I Am Your Sister" conference in Boston in 1990 to honor Lorde. But even this special edition, with its tributes by fellow writers and activists, including Adrienne Rich, Jewelle Gomez, and Barbara Smith, does not render in words Lorde's final battle with cancer. Instead, the tributes describe why Lorde and her work mattered, and still matter, and, as Gomez puts it, "the ways in which she used her words to point toward new paths, new ways of choosing to be in the world" (1997a, 84). Although these artists and activists who praise her are in mourning, they are not melancholic; they seek to celebrate the continuation of Lorde's voice in her writing. As Rich reminds us, "Audre Lorde was a great poet because she understood the necessity of poetry to our collective life and because she knew that beauty and bearing witness to the harsh materials of human struggle need never contradict each other" (82).[21]

These women who have come to praise Lorde acknowledge both her acceptance of and her resistance to the isolation of dying. Her last poems, posthumously published in the collection *The Marvelous Arithmetics of Distance* (1993), expose both her strength and her vulnerability in the face of death. Thus, the poem "Today Is Not the Day" begins with two stanzas in italics:

> *I can't just sit here*
> *staring death in her face*
> *blinking and asking for a new name*
> *by which to greet her*
>
> *I am not afraid to say*
> *unembellished*
> *I am dying*
> *but I do not want to do it*
> *looking the other way* (57)

According to Lorde, dying isn't simply something that happens to a person: dying is something one does, and, for Lorde, it must be done with all one's attention, not "looking the other way." Lorde begins her poem, therefore, with the "unembellished" certainty that she is dying, but also with another certainty with which she opens the third stanza: "Today is not the day. / It could be / but it is not" (57). The poem is testament to this knowledge of life in the midst of dying. It tells not what today is not, but what today is, including labor and loving and the presence of others: her daughter, son, and lover. Lorde's poem presents the gap between "today" and "not the day," between being and nonbeing, between her presence amongst the living and her imminent death. Death is close, and yet in an effort to protect her loved ones—but perhaps not to protect her readers after her death?—from the dangers of its proximity, her poem is meant to reassure: "I dare not tremble for them / only pray laughter comes often enough / to soften the edge" (58).

In *A Litany for Survival*, a film about Lorde's life and work, we become witnesses to her attempt to resist the isolating effects of illness. Although she acknowledges to her daughter, Beth, the need to leave New York and to live in the healthier environs of St. Croix, at the same time as we see her body and hear her voice weakening from cancer, she insists to her partner, Gloria Joseph, that she must continue to make connections beyond herself. But, she also admits that cancer, in transforming her voice, has transformed her relationship to herself. Hoarse from the harsh cancer treatments, her voice is no longer familiar, and the words she writes and reads back to herself are also no longer familiar, are, indeed, uncanny. Part of the process of transforming silence into language and action is recognizing that once her words are spoken, they are no longer her own. In this case, they are no longer her own even to her own ears. And, yet, she knows "it is better to speak remembering we were never meant to survive." For someone who was never meant to survive, to be able to say "today is not the day" is a political act.

In Lorde's work, and in the work of others about her, we have her death alluded to but not described explicitly. In *Cancer in Two Voices* and *Before I Say Goodbye*, the deaths of Rosenblum and Picardie are also framed by the representations of others. Rosenblum's death and the mourning that follows it is described in the penultimate chapter of *Cancer in Two Voices* entitled "The Last Sabbath February 1988." This chapter opens with Butler writing ten days before Rosenblum's death about Rosenblum's weekly discussions with Rabbi Kahn concerning Jewish spirituality and Rosenblum's relationship to it. Butler tells us that right up until her death Rosenblum was seeking new spiritual knowledge. When she is no longer able to meet the rabbi at his synagogue office, he agrees to pay a house call "to continue their ongoing dialogue" (1991, 167). When he arrives, Rosenblum asks Butler to join them and announces, simply, "Today we need to plan my funeral" (167). Butler recounts the discussion over what to include in the ceremony, and, in particular, Rosenblum's insistence that the service be "traditional" out of respect for her parents and their "need for ancient words, the prayers, the sounds" to comfort them in their grief.

After this entry about funeral preparations on February 4, we are presented with Rosenblum's last entry on February 7. She begins with the certainty that "I am now dying." No longer can she say, as did Lorde in her poem, that "today is not the day" because that would imply a future, which she understands she no longer has. "I am *now* dying" (168). Today is the day. Now. Dying. After this blunt beginning, she provides images in words of what *now dying* looks like. Remembering her younger self as an erstwhile photographer, Rosenblum writes, "Now I wonder what a series of photographs representing me dying would look like. How would I photograph myself now?" (169). She wonders how she would show the breakdown of her very being, and in response to this existential question, she gives us snapshots of herself at 9 a.m. at various intervals over the past three months. Her images record the reduction of her capacity to move out into the world. She describes the increasing stillness and isolation of a self—herself—being unmade. In the last image she describes, the only sign of life, the only break in the stillness of the scene is that her "left eyelid is lifted slightly, a sign that I understand that I am being photographed, but that I am too weak to speak or interact or perhaps even to care" (169). Her last words in *Cancer in Two Voices* are: "I am forty-four years old" (169). These last words, like the "now" in "I am now dying" confirm that there is no future or past in this present moment, only the utter stillness and simplicity of "I am forty-four years old." This now allows no future or past consolation. It is a moment beyond consolation.[22]

The next entry is from Butler on February 14, and it is the shortest entry in the book: "Valentine's Day. Barbara died today" (169). Between Rosenblum's last written words and her death, there is a silent gap of seven days followed by Butler's unelaborated statement. Butler's next entry after Rosenblum's death, on February 15, attempts to fill this gap with words describing Rosenblum's last days. It begins: "You lay on your bed, three days ago, swollen with edema, watching the doctor. 'You have a few weeks at most,' she said, an unexpected gift from this brusque woman" (170). It is unclear at this point what the "unexpected gift" is that is delivered in these words: is it the "few weeks more time" or the "at most"? And, it is also unclear who the gift is for—Butler or Rosenblum?

Butler continues the sequence of images of reduction that Rosenblum had begun, but could not finish, in her last entry: "The next day [February 13] your bloated, swollen body was unable to turn, to move, to walk. The mouth sores made speech uncomfortable. Reduced now from sentences to words, from glasses of ice water to a child's bottle, your hand too weak to hold the glass. 'I think it's time,' you whispered to me" (170). Knowing with "mournful clarity" that this might be one of the last nights she spends at Rosenblum's side, Butler keeps watch, offering reassurance that she will be by her side when death comes. One can't help but ask, again, who in this scene is in need of soothing reassurance? The next morning, having survived the night, Rosenblum sends Butler into the kitchen for fresh water—"smiling up at me with love," Butler tells us—and, when she returns moments later, Rosenblum is dead. Despite her vigil, Butler misses Rosenblum's death; and yet, even if she had been there at the precise moment, the question remains whether she would have captured the moment and been able to reproduce it in words as the final image, the final reduction, the moment without past or future, when being crosses over into nonbeing.[23]

Butler ends this chapter describing the funeral ceremony, planned by Rosenblum, and the week of sitting shiva, the Jewish period of mourning. At the funeral, during the reading of the Kaddish, the prayer for the dead, Rosenblum's mother, Regina, breaks down and wails in grief:

> Her wailing rose above the murmuring of the assembled, gathering force, gathering strength. As her throat pushed forth these primitive sounds, I remembered the sounds of labor and birth, the elemental sounds the body emits when it is beyond its capacity to tolerate pain. Her wails became keening and the keening blurred back into language, pleading. Words she was speaking to Barbara, to God, to anyone who could erase this moment. (171–72)

What Butler describes here is similar to those experiences that Elaine Scarry describes in *The Body in Pain*, in which language is destroyed and reduced to what Scarry calls "pre-language." In her phenomenological analysis of pain, Scarry, like Butler above, does not stop at the destruction of language that comes with pain. Rather, she insists as well that "conversely, to be present when a person moves up out of that pre-language and projects the facts of sentience into speech is almost to have been permitted to be present at the birth of language itself" (1985, 6).[24] The birth of language that we witness in the above scene does not, however, offer relief from the "elemental sounds the body emits when it is beyond the capacity to tolerate pain." Instead, what emerges out of the pre-language, the language of the body, is, paradoxically, a pleading for erasure.

Regina's mournful wailing and keening disrupts the consoling and redemptive account of Rosenblum's death Butler attempts to give. Butler does acknowledge that even Rosenblum, in all her "loving preparations," had "forgotten the necessity to leave space for the sound of the one who unexpectedly survives again. The sound a mother makes when she must outlive her first-born" (172). But, again seeking consolation, Butler must make Regina's grief fit into a day that is, according to her, "whole and intact" (172). In the end, Butler insists that Rosenblum "chose a conscious death" and "faced death squarely, honestly, saying everything" (174). But, Rosenblum's last words in *Cancer in Two Voices*—"I am forty-four years old"—seem to be about everything that cannot and will not be said about the future and, even, about now, *now dying*.

Just as it is up to Butler to record Rosenblum's last days, in *Before I Say Goodbye*, it is up to Picardie's partner, Matt Seaton. But, unlike Butler, he does not seek redemption in those last harrowing moments. And, Picardie, whose cancer, in traveling to her brain, has done precisely what she most feared it might do, is no longer able to cultivate an ironic self, nor any other sort of self. According to Seaton, on the night before she died, Picardie struggled up the stairs to help him put their children to bed for what would be the final time. He places their son, Joe, on Ruth's lap, and sets about giving their daughter, Lola, some milk. The bedtime ritual includes singing, usually led by Ruth, but now she is barely able to keep up with the rest of the family. Seaton relates the rest of this unbearable scene:

> After a few minutes, Joe simply got up off Ruth's lap, came over and lay down on the bed by me and Lola. The sight of Ruth's poor, hunched silhouette, half-lit by a shaft of light from the door, still faintly finishing the lines of our song, was the saddest thing I ever hope to see. I knew then that, like Eurydice, she was lost to the Underworld, and that the true meaning of dying is its absolute loneliness. (1998, 115)

As if to emphasize the absolute loneliness of dying, Seaton knows that this is only his experience of the scene, and not necessarily Picardie's experience at all. In fact, he cannot know what she feels in that moment because she cannot communicate to him in expression, gesture, or words: again, "the true meaning of dying is its absolute aloneness." The dying person "*falls* outside the *thinkable*," as Michel de Certeau says (1984, 190; emphasis in original), and, as if to signal this predicament, Picardie's writing ends abruptly in the middle of a column, which is then both finished and left unfinished by her sister, Justine. Her death is not given form in either her own words before her death or her sister's words after her death; her death is the failure of symbolization; it is the blank space on the page that writing can never fill. Neither of these texts, therefore, can present—and re-present—a picture of death itself, they can only provide a frame that attempts to give structure to what is lost. Seaton admits that he had imagined death as a "clean break between being and non-being" (103), but discovers instead that, for the person looking on as well as for the person who is terminally ill, "dying is a relentlessly attritional process of estrangement." Despite a desire to do and say the right thing, Seaton realizes "you are doomed to frustration, failure and regret" (105). Those who are left behind construct the frame, and both texts open and close with their words, not with the words of the person whose death is at the center of the story.[25] *And yet,* perhaps, there is an ethics to be found in such frustration, failure, and regret.

Language when faced with death must fail, but the failure of language is, according to Lacan, where the rest of the story must begin, the story of that which remains, the stories of those who remain (Certeau 1984, 192). What is called for, what is demanded in illness narratives, therefore, is, in the words of Cathy Caruth, a "new kind of listening, the witnessing, precisely, of *impossibility*" (1995, 10; emphasis in original). "But somehow," Picardie's sister, Justine, writes, "Ruth slipped away to a different place, a place where I could not go with her. It seems impossible: impossible to comprehend; impossible to find the words to describe the loss" (1998, 93). In the face of death, one fails. One confronts that which is impossible. One faces failure and impossibility as well as death. *And yet,* in illness narratives one "pays attention to suffering."[26] *And yet,* one attends to loss. *And yet,* one negotiates new terrains of embodiment and of language, of intersubjectivity and of utter aloneness. Through narrative, one moves out into the world and at the same time deeper into oneself. Failure cannot be overcome or compensated for, but it can be treated in and through narrative. What precisely is the function of this *and yet?* I will return to this question, as well as to the ways in which we might pay attention to suffering, in the chapters that follow.

4.

Becoming-Patient

Negotiating Healing, Desire, and Belonging in Doctors' Narratives

What has informed all of my discussions up to this point, sometimes explicitly and sometimes implicitly, is the emergence of HIV/AIDS in 1981 and the early days of the epidemic in the West, before the protease inhibitors came on the treatment scene in 1995. Although I take the publication of Lorde's *Cancer Journals* in 1980 as the moment of emergence of the figure of the politicized patient, it is impossible to imagine such an emergence occurring without the concomitant emergence of HIV/AIDS in the West. The figure and the genre emerged out of the ontological and epistemological rupture that was and is HIV/AIDS. There were and continue to be a multiplicity of responses to HIV/AIDS, and these responses—these medical, political, and aesthetic treatments—have transformed how we live now and, more particularly, how we do medicine, politics, and art.

In this chapter, I turn from the figure of the patient to the figure of the doctor both *as patient* and, alternatively, paraphrasing Deleuze and Guattari (1987), *as becoming-patient*. In order to explore the figure of the doctor as patient, I first take up *A Leg to Stand On* (1993 [1984]), a memoir by Oliver Sacks that relates his harrowing (for him) transformation, in the time before AIDS, from doctor into patient. From a discussion of doctor as patient, I move to a discussion of two more recent autobiographies, Abraham Verghese's *My Own Country* (1994) and Rafael Campo's *Poetry of Healing* (1997), which describe their experiences as doctors who treat people with AIDS (PWAs). I consider the possibility that doctors, especially AIDS doctors, might become-patients through desiring- and writing-productions. In order to clarify what I mean by this process of becoming-patient, I make use of the theoretical "pragmatics" of Deleuze and Guattari, and read these doctors' narratives of AIDS as describing, or utilizing Deleuze and Guattari's own terminology, as *drawing* a "range of potential" through desiring, healing, and writing, all of which are practices that encourage our "capacity to affect and be affected" (Massumi 1987, xvi).

Doctor as Patient

While hiking alone in Norway over twenty years ago, Oliver Sacks had a bizarre encounter with a bull, which led to an equally bizarre encounter with his own body. When he spotted a bull on a Norwegian mountain, Sacks, in terror, hurtled down the mountain and away from the bull. He took a nasty spill, and severely injured his left leg. Alone, miles from the nearest town or house, Sacks was certain he would die of exposure if he stayed on the mountain for even one night. And so he fashioned a splint out of his umbrella, and began a crab-like crawl down the mountain. He was eventually—and rather miraculously—rescued at nightfall by a hunter and his son, still far from his destination. As Sacks tells the reader, his rescue seemed to be the end of the story. "And yet," he writes, "there was to be another story or, perhaps, another act in the same strange complex drama, which I found utterly surprising and unexpected at the time and almost beyond my comprehension or belief" (1993, 22). It is precisely this other story that forms the ethical framework of *Treatments*. How do illness narratives (by patients but also by doctors) supplement the stories of disease, diagnosis, and cure articulated within the institutional and epistemological frameworks of medicine? At the same time, how do illness narratives inevitably fail to articulate the whole story of illness?

For Sacks, the other story that he must tell is a neurological and phenomenological story of "a fundamental disruption and dissolution of the leg, a disruption and dissolution at once physiological and existential" (108).[1] As Sacks soon discovers, there is some neurological damage to his leg, and as a result, he experiences agnosia, the inability to recognize his leg as a part of his own body. Sacks has the opportunity to explore neurological as well as phenomenological questions via his own body, in a situation not unlike that of Henry Head, the English neurologist who had the nerves in his wrist severed to see what would happen, though of course Head's decision to explore neurological phenomenon through his own body was voluntary while Sacks's was not.[2] In *A Leg to Stand On*, Sacks moves from bodily alienation, to the passivity of patienthood, to his eventual recovery and "restitution" to full "morality."[3]

Sacks's perception of his left leg as an object, as not part of his body—as silent and dead—affects not only the leg as a localized entity apart from the rest of his body, but also his overall body-schema and his sense of *being* itself. As Sacks writes,

> what was now becoming frightfully, even luridly, clear was that whatever had happened was not just local, peripheral, superficial—the terrible silence, the forgetting, the inability to call or recall—*this* was radical, central, fundamental. What seemed at first to be no more than a local, peripheral breakage and breakdown now showed itself in a different, and quite terrible, light—as a breakdown of memory, of thinking, of will—*not just a lesion in my muscle, but a lesion in me*. (1993, 46; emphasis in the original)

Sacks does not know his own leg; it is "utterly strange, not-mine, unfamiliar" (51). It is not *his* leg[4]—*that* is lost to him—but a "cylinder of chalk," an object that, nonetheless, is attached to him, and so disturbing to his body image. Sacks's description of his leg as a "cylinder of chalk" is reminiscent of Merleau-Ponty's discussion of agnosias and phantom limbs in *Phenomenology of Perception* (1962). According to Merleau-Ponty, agnosias and phantom limbs are phenomena that force us to "find a means of linking the 'psychic' and the 'physiological,' the 'for-itself' and the 'in-itself,' to each other to form an articulate whole" (77). The "for-itself" of Sacks's leg has been lost, compromising his being as being "for-itself." That which distinguishes the being of entities from conscious being, according to phenomenology beginning with Husserl, is intentionality. Consciousness is defined as consciousness of something; consciousness is directed outward, forming an "intentional arc" surrounding the lived body in the world. In Sacks's case, this intentional arc is disturbed, and he must somehow recover what Merleau-Ponty calls an "organic thought through which the relation of the 'psychic' to the 'physiological' becomes conceivable" (77).

For Sacks, both his experience of bodily alienation through the apparent loss of his leg, and the experience of passivity associated with patienthood, "slackens the intentional threads" that connect his body to the world, and through which his body inhabits time and space (Merleau-Ponty 1962, xiii). "I aim at and perceive a world," Merleau-Ponty writes, and it is this aiming at, this intentionality, that is the "active meaning-giving operation which may be said to define consciousness" (xvi, xi). The body is not *positioned* in the world like an object, but *situated* in the world, which, for Merleau-Ponty, implies a dynamic "attitude directed towards a certain or possible task" (100). Although Husserl describes intentionality in terms of both its active and passive genesis, Heidegger and Merleau-Ponty—as does Sacks—tend to spotlight the active rather than the passive character of intentionality (Husserl 1960, 77–79).[5] Thus, Sacks views becoming a patient as an "abdication of activity" as well as a "moral" reduction of self.

Although illness and disability cause a shift in the character of one's intentionality, I would contend that they are not conditions in which there is a *lack* of intentionality. Rather, in illness and disability, active intentionality seems to shift to the background, to become the spatial and temporal horizon of one's relatively passive physical being-in-the-world. Elsewhere I have written on the question of passivity and intentionality in relation to the case of Jean-Dominique Bauby, who suffered from locked-in syndrome and who, more than Sacks, experienced such extreme passivity that perhaps we can speak of, as Levinas does, something "beyond intentionality," where consciousness is both outward- and inward-directed (Diedrich 2001; Levinas 1983). Nonetheless, Sacks's description of the experience of patienthood as a contraction of one's world and as a diminishment of one's intentional arc, as well as a diminishment of one's sense of oneself as a moral being is critical to the understanding of disability and illness as experiences that reveal the meaning of being-in-the-world.

From the moment Sacks begins his crab-like motion down the mountain in Norway, his very human-ness is compromised as he is no longer able to assume the taken-for-granted upright posture that is thought to be a characteristic of humans.[6] Sacks's morally compromised state continues as he assumes the role of patient and undergoes surgery and a lengthy rehabilitation. Sacks explains,

> There had been, for me—and perhaps there must be for all patients, for it is a condition of patienthood (though, one hopes, one which can be well- and not ill-handled)—two miseries, two afflictions, conjoined, yet distinct. One was the physical (and "physical existential") disability—the organically determined erosion of being and space. The other was "moral"—not quite an adequate word—associated with the reduced stationless status of a patient, and, in particular, conflict with and surrender to "them"—"them" being the surgeon, the whole system, the institution—a conflict with hateful and even paranoid tones, which added to the severe, yet neutral, physical affliction a far less tolerable, because irresolvable, moral affliction. I had felt not only physically but morally prostrate—unable to stand up, stand morally before "them," in particular, before the surgeon. (1993, 129)

For a time at least, Sacks experiences the doctor-patient relationship from the patient's side of the binary, and this experience of patienthood requires that he rethink his profession as a neurologist. As Thomas Couser notes, "Sacks believes his accident has put him literally in a position to correct the scotoma—blind spot—at the heart of neurology, its ignorance of the patient's experience, its willed and sometimes callous objectivity" (1997, 188). But what is perhaps most important to Sacks's story is not his experience of patienthood, but his recovery, a recovery that involves, ultimately, the ability to "stand up" both literally and figuratively, both physically and morally. When his leg is restored to its full function, his status—that is, his human status and his status as neurologist—is restored as well. In a sense, then, Sacks returns to his rightful position (at least to him) in the doctor-patient binary. Sacks's narrative, according to Couser, "is a single-experience narrative written from the safe side of the border between disability and 'normality.' Whatever his intentions in this book, his brush with disability seems to have led him unconsciously to distance himself from a condition he found thoroughly alien and threatening" (187). Because he cannot imagine his moral affliction resolved without resolving his physical affliction first, Sacks does not ultimately transform the position of the patient within medicine, nor for that matter does he transform the position of the doctor. And, furthermore, the reason Sacks gets to tell this narrative of the physical and psychological estrangement of disability is because of his status as a doctor, not because of his reduced status as a patient.

Sacks's final chapter is entitled "Understanding," and the understanding he arrives at is both neurological and phenomenological. In *Phenomenology of Perception*, Merleau-Ponty notes that "[t]o understand is to experience the harmony between what we aim at and what is given, between the intention and the performance" (1962, 144). It is the return of this harmony that begins Sacks's recovery and understanding. And, for Sacks, this harmony is quite literally found in the "musicality of motion," or what he calls his "kinetic melody" (1993, 119). The body's modality toward the world is most often taken for granted. This means that one does not inhabit time and space in a self-conscious way;[7] one's gestures towards the world—from simple to complex—are not "a collection of partial movements strung laboriously together," but are composed like a melody (Merleau-Ponty 1962, 105). By listening to Mendelssohn's Violin Concerto, Sacks is able to *remember* "walking's natural, unconscious rhythm and melody" (1993, 117). "I felt," Sacks writes, "in those first heavenly bars of music, as if the animating and creative principle of the whole world was revealed, that life itself was music, or consubstantial with music; that our living moving flesh, itself, was 'solid' music—music made fleshy, substantial, corporeal" (94). Sacks's dis-ability—the loss of his "kinetic melody"—is, thus, only temporary.

In some respects Sacks's story is somewhat anomalous to the larger story that I am telling in *Treatments*, which has to do with the fact that illness in general and terminal illness in particular is precisely about the relationship of the social and individual self to loss. In my last case concerning breast cancer narratives, I highlighted that one particular mode of being ill, which I identified as a specifically American mode, attempted to transform loss through illness (and disability) into gain. But, what I did not consider in that case, because my examples were both terminal examples, was the *reality* of getting better (as opposed to the *rhetoric* of being better because of illness). Clearly, much of the time, illness leads to recovery not death, and many illness narratives, as Frank for one acknowledges, feature recovery rather than inexorable decline or catastrophic breakdown. If Sacks's story is something of an anomaly in my discussion here, nonetheless, it is not an anomalous illness narrative. It is useful as an example of a narrative in which loss can be viewed from the outside. As both patient and doctor, disabled person and non-disabled person, Sacks articulates loss from both sides of the divide, a divide that is an ontological, epistemological, and ethical one. I do not mean to imply, however, that Sacks is in a rarefied position to tell the whole story of loss (though he seems to think he is). Rather, his story suggests the multiple positions in relation to loss and the telling of loss. Despite, or more likely because of, the elusiveness of loss, I want to think about, in and through my treatments of illness narratives, the possibility of holding onto loss: loss not as that which is understood, and from which one recovers, nor as that which must and can be converted into gain, but *loss as such*. Thus, a question that Sacks can't answer (nor does he even bother to ask), but that informs this project is: what if the loss itself—or the losing—characterizes one's being-in-the-world? Loss as such will characterize the experience of HIV/AIDS from both sides of the doctor-patient binary in the 1980s.

Becoming-Patient

Although Sacks is confronted with the possibility that medicine might fail him, ultimately he and medicine recover their status. But, unlike the experience of temporary disability that Sacks describes, in the case of HIV/AIDS in the 1980s,[8] the failures in vision, knowledge, and practices in medicine are not temporary setbacks, but seem to present a more permanent crisis.

In order to further elucidate these three failures, I turn now to two autobiographies written in the 1990s: Abraham Verghese's *My Own Country* (1994) and Rafael Campo's *Poetry of Healing: A Doctor's Education in Empathy, Identity, and Desire* (1997), which describe the effects of treating HIV/AIDS on each doctor's identity, on his desire for community and belonging, and on his identification and/or disidentification with the medical profession in the United States.[9] Campo is a gay Cuban-American who grew up emotionally exiled and disidentified from both his Cuban ethnicity and his gay sexuality. Educated at Amherst and Harvard Medical School, he eventually finds himself working at large public hospitals in San Francisco and Boston where he encounters numerous people with AIDS. Verghese is an Indian raised in Africa and educated in India and the United States who works as a specialist in infectious diseases in a small Tennessee community. He becomes, unexpectedly, thanks to an "uncanny convergence" (Grosz 1999, 3) of a doctor and a community, that community's "AIDS expert."

Campo and Verghese see themselves as insiders and outsiders within the medical profession in the United States, and within the communities in which they practice medicine. They both challenge the doctor-patient dyad that is a cornerstone of modern, Western medicine, and they do so not only in the ways they practice medicine, but also in the ways that they bring the body into language through their writing. Campo and Verghese, in different ways as we shall see, provide narratives that blur the rigid boundaries between science and literature. Their narratives seek to supplement the doctor's more usual discourse exemplified by the case history and medical chart with what Campo calls "the poetry of healing." Both writer-doctors make an explicit connection between the practices of writing and healing, and a distinction between the doctor who is concerned only with curing as a heroic activity and the doctor who is always also "a witness to another man's ailing body" (Campo 1997, 13).

My readings of Campo and Verghese revolve around three key terms provided by Campo's subtitle—identity, empathy, and desire—which I read through the theoretical "pragmatics" of Deleuze and Guattari, as well as the theories of those, particularly feminist philosophers Rosi Braidotti and Elspeth Probyn, who take up Deleuze and Guattari's "tool-kit" in order to ask new questions about the ways in which identity might be conceived as nomadic, empathy might be achieved by becoming-other (or what I call becoming-patient), and desire might be perceived as productive, creating lines of flight and other futures. The other term in Campo's subtitle—a "doctor's *education*"—is also important when viewed in relation to Deleuze and

Guattari's theoretical pragmatics. Both Campo and Verghese offer stories of doctors' educations, but these are not conventional stories of medical school and the triumph of the Western medical knowledge. In fact, their medical school education does not provide them with the tools they need in their encounters with PWAs (person with AIDS).

Arthur Kleinman, whose book *The Illness Narratives* explores the impact of chronic illness on how medicine is practiced, insists that doctors must not simply pay attention to disease, but to the "symbolic network linking body, self, and society" (1988, 6). Following George Engel (1977), Kleinman calls this "alternative therapeutic approach" the "biopsychosocial model"[10] of illness that calls for a "reconceptualization of medical care as . . . empathic witnessing of the existential experience of suffering and . . . practical coping with the major psychosocial crises that constitute the menacing chronicity of that experience" (10). Although Kleinman's work deals mostly with chronic ailments—such as chronic pain, chronic fatigue syndrome, and depression—and although it would be hard to describe HIV/AIDS as a chronic illness at the time Campo and Verghese are writing,[11] nevertheless, HIV/AIDS forces both doctors to imagine and practice "alternative therapeutic approaches" that allow them to be doctors interested in disease, as well as empathic witnesses interested in the "remoralization" of their patients (54), just as Sacks was interested in his own remoralization. The sort of education required, then, is not in how to better "construct a case report" as an official account with "legal and bureaucratic significance" (131). What is required is an *unlearning*, as Kleinman understands it: "a return to the time when as beginning medical students, with a foot in both lay and professional worlds, [practitioners] audited the speech of their first patients with great intensity, with something approaching awe in respect for hearing the patient's story in his or her own words and with deep sympathy for the human condition of suffering" (130). To some extent, such a return requires a loss of status, a loss of ego, a loss that is welcomed, even perhaps sought.

Doctor/Outsider[12]

My Own Country opens with a return home: a young man journeys South from New York City to visit his parents in Johnson City, Tennessee. His journey is both familiar—his route is well traveled and well known to him—and unfamiliar: he recognizes the external landscape, but the landscape of his own body—now ravaged with illness and fear of what his illness portends—is somehow now unfamiliar. This change in the internal landscape of his body changes his relationship to the external landscape surrounding him. "His breath is labored," Verghese writes, "weighted by fear and perhaps by the knowledge of the burden he is bringing to his parents. Maybe he thinks about taking the next exit off Interstate 81 and seeking help. But he knows that no one can help him, and the dread of finding himself sick and alone keeps him going. That and the desire for home" (1994, 9). His journey ultimately ends in the ICU at the Johnson City Medical Center, where, according to Verghese, "[t]he hometown boy was now regarded as an alien," the first documented case of AIDS in Johnson City (13). The case with which Verghese opens his narrative occurs before his own arrival in Johnson City, and he represents these events as he imagines they might have happened. He imagines as well their particular significance to the young man, his family, and the medical personnel at the Johnson City Medical Center. In this opening scenario, Verghese creates a fiction of return that will be significant to the rest of his story.[13]

After his own arrival in Johnson City, Verghese himself soon discovers cases of HIV/AIDS in the small, Southern town that should have been, according to early epidemiological studies that Verghese has read, an "unlikely" place for HIV. Indeed, he soon discovers that HIV/AIDS does not remain within the imaginary boundaries of various large American cities, but instead migrates quite freely along the interstate highway system; most often in the bodies of gay men, who go from small towns like Johnson City to the large cities of New York, Washington, D.C., and Atlanta in search of a place to articulate their gay identities, and sometimes also in the bodies of truckers, who have sex with male and female prostitutes along their routes. Eventually, Verghese will map the peregrinations of the persons he treats for HIV/AIDS. From his patients' medical facts as well as their personal stories, Verghese is able to identify a paradigm: "The first two steps of the paradigm—leaving home and then the period of urban living—were followed by the long voyage back" (323).[14] The voyage out is marked by a deterritorializing desire for other individuations and other pleasures, while the voyage back is marked by a reterritorializing desire for the hoped-for safety of home and a nostalgic past as a conservative (in the sense of conserving energy, not necessarily tradition) response in the face of inevitable death.

The desire for home that motivates many of Verghese's AIDS patients is a desire that motivates him as well. Like the AIDS virus itself, Verghese arrives in Johnson City having migrated far from his origins. He notes that his own arrival in America as a "rookie doctor" in 1980 roughly coincides with the arrival of the AIDS virus in the U.S. port cities of New York, San Francisco, and Los Angeles (16). *My Own Country*, then, is not only the story of the migration of a deadly disease, but also the story of the migration of an individual, his family, and his "race." According to Verghese,

> In their herald migration, my parents individually and then together reenacted the peregrination of an entire race. Like ontogeny repeating phylogeny—the gills and one-chamber heart of a human fetus in the first trimester reenacting man's evolution from amphibians—they presaged their own subsequent wanderings and those of their children. (18)

Migration—or "nomadism"[15]—is an exemplar condition of postmodernity in which national, racial, and bodily boundaries are discovered to be neither impermeable nor fixed. Verghese acknowledges his own "nomadic subjectivity," in Rosi Braidotti's terminology, in his habit of reinventing himself in each new place, and in what he calls his "chameleon like adaptability" (51). He realizes that in Johnson City he is a "messenger from another world," and that his foreignness, paradoxically, allows his patients to reveal to him intimate details of their lives without fear of his judgment. Both his adaptability and his foreignness are useful in his medical career in general and in his HIV/AIDS practice in particular.[16] Yet, despite his recognition of the benefits of his own nomadic subjectivity and despite his sense that "a long voyage was over" and that "this corner of east Tennessee" was his "own country" (41), nonetheless, Verghese is haunted by a feeling of rootlessness, and he yearns for a permanent home to return to.

Like Verghese, Campo also finds it difficult to locate the community to which he finally and fully belongs. Because of his homosexuality, he never feels fully a part of the predominately conservative and Catholic Cuban-American community, and, because of his ethnicity, he never feels fully a part of the mainly white homosexual community. He is, in his own words, "doubly illegal" (1997, 263). His choice of medicine as a profession is an attempt to cancel out if at all possible this double illegality that he senses defines him. "[I]t was during my adolescence," Campo writes, "as my body began to speak more boldly its language of desire to the outside world, that I also began to understand medicine as a 'desirable' profession." Campo continues, "As a child of immigrants, I imagined that my white coat might make up for, possibly even purify, my nonwhite skin; learning the medical jargon might be the ultimate refutation of any questions about what my first language had been" (18). For Campo, his doctor's identity provides a measure of immunity and a definition of self.

Yet, the white coat and the M.D. after his name designating him as "disengaged, sexless, AIDS-free, and possibly straight" (140) are not potent enough to assuage his desire and keep him immune from his own becoming-queer. Medicine gives him an identity free of pathology, but it also provides him with a way to pathologize himself: "a way to confirm what it had been telling me in other ways since I had first considered a career in medicine: that I did not belong, and that I was not welcome. That I was *unhealthy*" (183; emphasis in original). Ultimately, however, it is Campo's own desire and love for another man, a desire that he had attempted to fortify himself against, that leads him to a kind of belonging. As Campo explains,

> Ironically, it was this unspeakable love that ultimately led me to the place where I am now, to a career not in diagnostic radiology but in general internal medicine, which allows me to provide primary care mostly to Latino patients in Boston. It was the love of another man, Latino himself, who taught me to love my culture, which led me to my place at the banquet table. (263–64)

His desire leads to unspeakable love, which leads eventually to other, unexpected places and other, unexpected futures. M.D., Campo discovers, stands not only for "Medical Doctor" but also for "My Desire."

For Campo, in the act of desiring, one belongs; desire produces connections and creates belonging. While Verghese desires *for* home, and sees a fixed and stable home as a culmination of his desire, Campo understands desire itself *as* home. Verghese wishes that his desire for home would be consummated once and for all, but Campo knows belonging is found in the *act of desiring*, or, the "movement of desiring belonging," according to Elspeth Probyn. In her work, Probyn is interested in "the inbetweenness of belonging, of belonging not in some deep authentic way but belonging in constant movement, modes of belonging as surface shifts" (1996, 19). It seems to me that Campo, like Probyn, conceptualizes belonging in terms of desire, movement, and inbetweenness, while Verghese is haunted by a nostalgia for "some deep authentic way of belonging" that he has never, in fact, experienced. We might say, then, that Verghese emphasizes the *being* in belonging, while Campo "emphasizes [the] *longing* in belonging" (Probyn 1996, 13; my emphasis).

Empathy: Breaking Down the Doctor-Patient Binary

In the age of AIDS, where sexuality has become linked with death in the popular imagination, desire has had to be rethought. Campo rethinks desire in surprising ways; for him, desire spans the chasm between doctor and patient (traversing in both directions) and facilitates empathy and healing. And so *The Poetry of Healing* opens unexpectedly with a patient's erection, and with Campo's recognition that the encounter between doctor and patient is one that produces desire, despite the fact that, within the institution of medicine (a Deleuzian "plane of organization"), such a desire is denied through practices, instruments, and spaces that create distance rather than proximity between doctors and patients.[17] In the case of AIDS, at least in the early years of the epidemic, the doctor-patient binary was breached, and the hierarchy between the two was flattened, substituting, if only for a time, a "rhizomatic model" of relationship for an "arborescent model."[18] In what ways, then, do Verghese and Campo have to *unlearn* much of what they learned in medical school in order to offer treatment for HIV/AIDS: treatment, in the rare sense of the word, as "discussion or arrangement of terms, negotiation"?[19]

The reluctance to mount a full-scale effort against the AIDS epidemic on the part of U.S. government officials and health professionals resulted in the gay community's creation of an HIV/AIDS care network to take care of its own. As a consequence of the epidemic in general and this care network in particular, according to Tamsin Wilton, "the entire concept of medical expertise was momentarily stood on its head" (1992, 103).[20] PWAs and their caretakers became as expert as, if not more expert than, their doctors on the nature of HIV infection and their treatment options. As I discussed in chapter 2, what occurred in the case of AIDS was a "politicization of patienthood," as patients began to demand that government in general and individual doctors and researchers in particular respond to their needs for appropriate and empathic health care (Singer 1993, 105; Epstein 1995). Thus, in *My Own Country* Verghese describes not only his own, but also his patients' transformation into "AIDS experts" and activists. For example, one of his patients, Fred Goldman, is instrumental in forming the Tri-City AIDS Project (TAP), a community-based group in Johnson City that disseminates information, organizes a support group, and provides countless other services for PWAs. Verghese understands that Goldman, thanks to his activist past, has "spent a lifetime training for the role of AIDS activist in Johnson City. The stakes had never been higher; his activism now had a clear focus. It would be his therapy, it would give him long life, it would greatly better the lives of persons infected with HIV in upper east Tennessee" (157).

Another of Verghese's patients, Vickie McCray, who contracted HIV from her husband, Clyde, who had been having sex with men without her knowledge, also becomes active in TAP. Despite Clyde's betrayal, Vickie cares for him at home until his death. Eventually, when Hospice of Johnson City is brought in to help Vickie provide care at home, she quickly learns new knowledge and skills, Verghese tells us, to take Clyde's "vital signs, recording in a spiral bound notebook what transpired on her shift: how much fluid he drank, how many times he urinated, his level of consciousness, and his blood pressure. She had even picked up on words like 'diaphonetic' and 'dyspneic'" (285–86). Vickie tells Verghese that she believes she has been given AIDS for a reason, that it has been the catalyst in transforming her from a woman whose life revolved entirely around her family to a woman whose closest friends are gay men, who is on the board of TAP and enrolled in nursing school. Without a trace of irony, Verghese says Vickie's transformation "was indeed a TV movie sort of transformation" (334).[21] Vickie, or at least Verghese's description of a somewhat beatified Vickie, exemplifies an art of being ill and doing illness, as I described in the previous chapter, that emphasizes the ways in which illness is an opportunity to live better, to become a better person. Her infection doesn't make her ill, at least not during the time Verghese is in Johnson City, rather it actually seems to improve her life by motivating her to do things she never imagined doing before.

Although somewhat less sentimental than Verghese, at least initially, Campo too is inspired by and finds meaning in the lives of his patients who are facing death. One in particular, a cancer patient named Mary, "smoothly bald and pale white from chemotherapy," is a patient Campo comes to love, despite the fact that they are constrained by their "respective worldly roles, as doctor and patient, gay Latino man and straight white woman" (1997, 26). Campo describes their love as transcending gender and sexuality:

> Highly erotic and deeply pleasuring without our ever having slept together, as commonplace and yet unexpected as life crossing over to death, immortal as each retelling or the act of writing. Both Mary and I left our loving friendship healthier, I think, closer to being cured. She waved to me as she left the hospital, still bald, still beautiful, but more full of life, the life we shared. (26)

Campo's description of himself and Mary—doctor as well as patient—as "healthier" and "closer to being cured" also reiterates an art of being ill and doing illness concerned with improvement. According to Campo, one might become cured not in reality but in effect (for only Campo survives Mary's cancer) if one behaves a certain way: that is, like Mary, who becomes "more naked, more available and accessible to others, more beautifully free" (25). Although Verghese himself is not American, and Campo, in terms of his ethnicity and his sexuality, is not a "typical American" (whatever that might be), they both tell typically American stories. Verghese's stories of Vickie and Fred and Campo's story of Mary are stories told for the self. Below I will consider for whose self—the patient's or the doctor's—these stories are told.

Although Verghese describes Fred and Vickie as transformed into heroic "generals" leading the battle charge against AIDS, his own transformation into an "AIDS expert" eventually becomes a burden he struggles to bear. Verghese chronicles the early years of the AIDS epidemic in this country (specifically between 1983 and 1986), and admits that, "[i]n the early days, dealing with AIDS made us an elite group, an unexpectedly glamorous group. Even the cardiac surgeons could not approach our kind of heroism. Yes, they dealt with death everyday. But it was somebody else's death they had to worry about. Never their own" (67). Although not entirely seduced by the prevailing image of the omnipotent and omniscient twentieth-century Western medical doctor, Verghese admits that infectious diseases had been an appealing specialty for him because it was "the one discipline where cure was common." According to Verghese, "In the battle of man against microbe, man was winning. Astute diagnosis was rewarded by a return to perfect health" (25). Thus, when AIDS arrived on the scene, Verghese, along with most other medical professionals, believed that it would be necessary simply to "transfix it with our divining needles, lyse it with our potions, swallow it and digest it in the great vats of eighties technology" (24).[22]

It wouldn't take long, however, to discover that in the battle between "man" and the AIDS "microbe," the microbe was winning. Medicine's divining needles and potions were not working, and, unable to rely on established techniques of doctoring, Verghese's forms of treatment radically change. "In the absence of a magic potion to cure AIDS," Verghese writes, "my job was to minister to the patient's soul, his psyche, pay attention to his family and his social situation. I would have to make more home visits, make more attempts to understand the person I saw in the clinic, be sure I understood the family dynamics by meeting all its members" (219). But, as Verghese acknowledges, his training had not prepared him to be this kind of doctor:

> We were trained in hospitals, not in patients' homes; we were biased toward technological interventions in the form of drugs and needles; words like the "soul," the "spirit," were considered dirty words. There was no or little payment for the nontechnological kind of medicine: hand-holding, family visits, home visits had no billing codes; bronchoscopies, colonoscopies and PAP smears did. For me to practice the folksy kind of medicine that was

> required would not only cost me monetarily, it came at the expense of time with my family and time for myself. (219)

And, despite the fact, or perhaps because, he is so adept at practicing the art of medicine, Verghese eventually experiences extreme melancholia brought on by so much death around him and so little in the way of remedy: in short, he becomes burned out.

After the death of one of his AIDS patients, Ed Maupin, Verghese notices a distinct change in the "shades and colors" of his life:

> Now it seemed as if everything I witnessed was imbued with this sense of loss. I was a doctor, a scientist, trained in professional detachment, but all the usual postures seemed satirical in the face of AIDS. I felt these deaths. I was filled with a longing for home (whatever I conceived that to be) so strong that I sometimes wondered if I myself was dying and this feeling was foreboding, the bittersweet messenger. (185)

The longing for any means of treating this deadly disease is intimately connected for Verghese with a longing for home, and the failure to have appropriate treatments to offer his patients is also a failure to satisfy his longing for home. Ultimately, empathy for him is not enough, and his battle is both an external one to acquire the means to fight the AIDS virus, as well as an internal one to keep his melancholia at bay. For the former battle, Verghese becomes his patients' "surrogate activist, their link to the larger consciousness of AIDS" (222); and, for the latter battle, he writes the story of an AIDS doctor and the patients he cares for. Doctor, activist, writer: these three commitments are intimately connected for Verghese.

For both Campo and Verghese, the interactions with PWAs that change the "shades and colors" of their lives are a manner of becoming, becoming through infection, which is not a becoming-infected *in reality*,[23] but rather a *desire for infection*, a desire that Verghese experiences as a combination of empathy and foreboding, and Campo experiences as a combination of empathy and loathing. Verghese's oscillation between empathy and foreboding seems to be about both the possibility of his becoming-infected by homosexual desire as well as the possibility of becoming-infected by HIV. His sleep and waking life are troubled by a recurring dream, which he calls his "infection dream."[24] "The 'infection' dream is so frequent," Verghese writes, "that many mornings I wake up and, even when I have shaved and showered, I am still having to convince myself that I am uninfected. Perhaps my insomnia of the last six months is an attempt to avoid the infection dream" (336–37).[25]

One of the many stories of private and public failure that *My Own Country* tells is the very personal failure of Verghese's marriage, which he relates at least indirectly to his obsession with his work as the AIDS doctor in Johnson City. His wife, Rajani, is unable to understand this obsession and fears for its consequences in terms of the health of the entire family, their position within the small, close-knit Indian community of Eastern Tennessee in which Rajani, at least, is firmly rooted, and also in terms of what it signifies about Verghese's sexuality. In Deleuze and Guattari's terminology, Rajani represents "the great molar powers of family, career, and conjugality," whereas Verghese is driven instead by "an irresistible deterritorialization that forestalls attempts at professional, conjugal, or Oedipal reterritorialization" (1987, 233). According to Verghese, "My work with AIDS in the community fell into [the] chasm between us. AIDS was like another wild friend, a friend from a different social stratum, a friend I indulged but no longer brought to the house or even discussed with her" (139). Verghese himself does not entirely understand his obsession with his AIDS work, or his intense curiosity about his patients' lives, but he readily admits that this curiosity has sexual undertones. Through his AIDS practice, Verghese gets close to a sexuality that had previously been unknown to him. Because of his proximity to homosexuality, Verghese is able to identify with its outlaw status. In no way do I mean to imply Verghese *is* gay; rather, I want to highlight the ways in which AIDS, in the West at least, is not simply about disease, but always also about sexuality.[26] As both Verghese and Campo make clear, doctors themselves are not immune to this implicit and explicit association. For Verghese, the association is sometimes thrilling, while for Campo it is terrifying.

Campo's oscillation between empathy and loathing (and this includes, importantly, self-loathing) is a result of his own sexuality and the link between homosexual desire and death that AIDS would come to signify in the West. It is not, then, the *possibility* of becoming-infected with homosexuality, but rather the reality of his becoming-queer that affects Campo's relationship to AIDS and to those PWAs he treats. For example, Campo resists getting close to his patient, Aurora, an HIV+ male-to-female post-operative transsexual, who embodies, literally, Deleuze's and Guattari's notion of becoming-woman and the endless process of desubjectification such a becoming requires. Campo initially only perceives Aurora through his "protective fiberglass-mesh mask" (1997, 30), shoring up the boundaries between them by busying himself with "collecting the data of her decline" (31). Campo uses science and technology as a screen between himself and Aurora. In the eyes of medicine, Aurora is a "freak," but it is precisely through her "freakery"[27] that Aurora teaches Campo about desire, and helps him to unlearn his own internalized homophobia and the techniques he uses to distance himself from his patients:

> Science failed to understand her, though it altered her body. Medicine did not love her, though it penetrated her with needles and x-rays. Only the act of writing can find her now, because it is the same journey she has made, from the imagined to the actual, from the transitory to the persistent. From the unspoken to this physical and loving lament. (32–33)

Like Verghese, Campo realizes that being a doctor, even the sort of doctor who attempts to become-patient, is not enough. In order to offer better treatments, he believes he must also become a writer; that is, he must create becomings through writing-productions. Faced with the failures of vision, knowledge, and practices that AIDS reveals within the institution of medicine, Verghese and Campo turn to writing as a treatment for themselves, their patients, and society in general. They attempt in narrative to overcome medicine's failure of vision by describing their patients as subjects who are ill, not as objects with a disease to be diagnosed; as individuals, not ideal disease types. They attempt to compensate for medicine's failure of knowledge by acknowledging what medicine does not and cannot know, and by supplementing expert knowledge with the everyday knowledge they learn from the stories of their patients' lives. Finally, they attempt to implement practices associated with not only the science of medicine but also the art of healing.

AIDS gives Campo, because of his own sexuality, more than a vague foreboding of his own death, it gives him an eerie certainty of his own death by this disease:

> In some ways, I know I have been dying of AIDS since the moment I first learned about the virus. Each smooth tube of blood I draw seems to come from my own scarred and indurated veins, each death note I have dictated has my own name and signature at the bottom of the page. Any disease that could erase from the world the bodies of so many people like me, people with whom I had not even had the chance to form the bonds of community, would seem necessarily to take with it small parts of my anatomy; AIDS has cut off the part of my tongue that once made it easy for me to sing, it has laser-ablated my seminal vesicles, it has occluded the blood flow to the area of my visual cortex capable of plainly seeing joy. (29)

Campo's empathy arises through the body and through desire, which he sees as "inherent in the doctor-patient relationship" (27). In a sense, then, in Campo's form of empathy, the body of the doctor displaces the body of his or her patient: in empathy, the doctor must always also become-patient, and this becoming can be enacted in writing (Deleuze and Guattari 1987, 240; see also Probyn 1996, 52). Writing provides a space within which Campo can explore what it might mean to be dying of AIDS. Because AIDS has affected so many people like him, Campo attempts to take his place along with these people, in language as much as in reality. And so, while Verghese seems to fear becoming-patient, Campo desires it.

Campo's desire to become-patient does not arise solely out of his perceived relationship with his AIDS patients and their queerness, but from his own experience with a pre-cancerous tumor in his wrist, the same wrist that he had injured in a skiing accident as a teenager. Because the ski outing had been arranged by an older man who Campo suspected, and also hoped, wanted to have sex with him, the wrist injury and the cancer are forever connected in his memory with the expression of desire and the potential consequences of desire. By becoming-patient, Campo realizes that his patients have gifts of healing to offer him too; they, in turn, are becoming-healers. When he has to tell one of his patients, a 74-year-old Mexican woman, that she has breast cancer, he is unable to find the words to tell her. When she reaches out and touches his left arm, where his own cancer had been cut from his body, in a gesture that communicated "deeply and wordlessly," Campo is able to rediscover his own narrative (226). Doctor, writer, cancer survivor, queer: these modalities all affect Campo's capacity to see and to think otherwise, and to practice a new kind of medicine.

For both Verghese and Campo, in this new kind of medicine, the practice of touch frequently becomes more effective than the invasive, heroic techniques associated with modern Western medicine. Elspeth Probyn has described her own work in cultural studies and queer theory as a "sociology of the skin," which requires, according to Probyn, "a heightened sensitivity to sensibilities, to be captured by other manners of being and desires for becoming-other that I call belonging" (1996, 5). Campo and Verghese, it seems to me, practice something similar: a *medicine of the skin*, perhaps. To a group of students, Verghese demonstrates this medicine of the skin when he "offers" to one of his dying AIDS patients the "ritual of the examination, the dance of a Western shaman" (343). Proceeding wordlessly, he saves his most potent means of examination for last. Through his unusual talent for percussion, Verghese is able to listen to the tissues and organs through a series of rhythmic touches on the surface of the body of his patient:

> I percuss his chest, and the sound of his right lung is disturbing. Only at the very top, near his collarbone, do I hear the *thoom* of resonance. Below that, from above his nipple to his belly, it is dull; the sound is indistinguishable

> from the *thunk* of the liver. The lung has been transformed from a spongy, light, pliant organ to a solidifying, consolidating mass. The sounds of my percussion on his body fill the room. *Thoom, thunk, thunk, thunk, tup, tup, tup*. I glance at his parents. They listen to the sound of their son as if mesmerized. Once more: *Thoom, thunk, thunk, tup, tup*—even Luther seems to pause in his delirious muttering, his floccillation, to listen to the music of his body, to relax, to smile. (343)

Unlike similar practices of examination and interrogation used to ascertain a pathological fact within the patient's body described by Foucault as emerging with the anatomo-clinical method,[28] the purpose of Verghese's examination is for more than its diagnostic significance; it has no signifier or signified; it is the touch itself that matters. In this moment, Verghese gives up interpretation in favor of an experimentation of touch across the binaries that separate doctor from patient. Verghese's examination produces, in Deleuzoguattarian terms, a body without organs[29] by "[t]earing the conscious away from the subject in order to make it a means of exploration, tearing the unconscious away from signifiance and interpretation in order to make it a veritable production," and, finally, by "tearing the body away from the organism" (Deleuze and Guattari 1987, 160).

Writing as Desiring-Production

It is not enough for Verghese to offer mere percussion as a gift to his patients. He feels compelled as well to practice a form of writing that isn't concerned with diagnosis, but with touching his patients across the gap that usually separates doctors from patients. He wants to give voice to the music of his patients' bodies, and feels compelled, as does Campo, to provide his patients with a space in which to tell their stories. It is important to note that the stories that both Verghese's and Campo's patients tell are not only stories of illness and death, they are also life stories in which illness is only a single facet, not the whole of it.

Although they are motivated by a similar desire to link writing with healing and a belief in the need to supplement the doctor's scientific and objective discourse with that of their patients' individual and subjective stories, the narratives Verghese and Campo offer are ultimately quite different, particularly in the ways they bring the body with AIDS to language. While on the one hand, Verghese attempts *to write about the body with AIDS*; on the other hand, Campo attempts *to write the body with AIDS itself*. This may seem to be a fine distinction, but it is particularly relevant if we consider whether the experience of AIDS demands a new form of writing.[30] The historical specificity of AIDS must be kept in mind—in terms of who writes about AIDS, why they write about it, and how they write about it. It seems likely that AIDS has opened a fissure in the myth of history as living and continuous, and has disturbed the project of modernity and the march of progress. How might one begin to describe this fissure, this rupture, and what purpose might such a description serve?[31]

For Verghese, the purpose of his writing is not so much to describe the rupture that is AIDS in its positivity, but to offer words, to tell stories—the stories of his patients—as the means to repair it. *My Own Country* is, finally, a reparative narrative, or, perhaps, in Deleuze and Guattari's terminology, a reterritorializing narrative; though, it seems to me, that it is most notable, paradoxically, for its failure to repair or reterritorialize. Again, Verghese's failure is both a failure to heal his patients and a failure to belong, because healing and not healing, belonging and not belonging forever overlap for Verghese. And so, he ends *My Own Country* with a return to the narrative with which he began: the journey home of the first AIDS patient in Johnson City. But, six years on, it is a different Johnson City and another future, and Verghese realizes, "if his voyage were to happen today, he might find a community in Johnson City better equipped to deal with him, to accept him. I have faith in the town and its people" (347). Ironically, however, Verghese acknowledges his faith in the town and its people as he and his family journey by car away from Johnson City to a new life in Iowa. It is this movement from Johnson City that opens up the conditions of possibility of belonging, a belonging, paradoxically, born in departure. In tracing this line of flight from Johnson City to Iowa and, later, to El Paso, Texas, Verghese replaces his nostalgia for a past that never was with a nostalgia that is productive of future possibilities of belonging. No longer does Verghese understand belonging as about a lost past; now belonging is about a future to be found or invented. In *Outside Belongings*, Probyn describes this sort of nostalgia, a nostalgia that, for her, becomes genealogy: "As it moves between present and past, nostalgia is no longer tied to an origin or a cause. Rather, like desire, it produces its object. It scrambles any sense of a fixed beginning" (1996, 116). Verghese uproots himself and his family from Johnson City to better cultivate his skills as both doctor and writer. By leaving Johnson City and distancing himself from the story he needs to tell, he is finally able to write about it; and in his words that describe the impossibility of belonging, Verghese discovers a kind of belonging, a desiring belonging.

While Verghese requires distance to bring his own desiring belonging to language, Campo requires proximity, a proximity in which words and bodies collide on the surface of a desire that, Campo writes, "I have always known belongs to all of us" (1997, 14). His opening chapter provides a diagram of his project and a diagnosis of *healing as desiring-production*, as opposed to *medicine as social-production*.[32] His narrative moves from his patient's body to his patient's desire to the act of writing as an encounter with "the shocking, empowering energy of a great desire" to his own body, a doctor's body. Patient-desiring-writing-doctor; that is, desiring and writing as bringing bodies—the bodies of patients and doctors—together on the same plane. In a moment that reveals both desire and its danger, Campo is pricked by the same needle he has just placed in the veins of one of his AIDS patients, who emerges from a kind of oblivion into life, his spewing blood covering Campo. According to Campo, "When the same needle pierced my own skin, my first thought was to deny the literal connection between us" (58). To deny the literal connection between doctor and patient is what modern medicine in its usual guise attempts to do, but with regards to AIDS at least, as Campo and Verghese show, modern medicine also fails to deny this connection. Medicine's failure is a creative failure, however; it is a failure that leads to strange becomings.

In that moment of being pierced by the blood of another, a moment of vulnerability, not omniscience or omnipotence, Campo writes, "I finally knew how human I was, I was made acutely aware in one terrible moment that all any of us has in the world is the same body" (60). "After a real rupture," Deleuze and Guattari write, "one succeeds . . . in being just like everybody else" (1987, 279 ellipses in original): the same as every*body*, *all the same body*. According to Donna Haraway, the postmodern body is a "construct of always vulnerable and contingent individuality" (1991, 220–21), and can be distinguished from the modern body, characterized in modern biomedical discourse as autonomous and "fully-defended" (224). "[T]he most reliable Western individuated bodies," Haraway insists, echoing Probyn, "neither stop nor start at the skin, which is itself something of a teeming jungle threatening illicit fusions, especially from the perspective of a scanning electron microscope" (215).

The Rhetorics of Suffering

The above discussion of Campo and Verghese accepts that writer-doctors can tell the stories of their patients; it accepts that such tellings are both necessary and possible. Now, however, I want to consider another possibility: that such narratives are examples not of doctors becoming-patient through their own desubjectification, but are, rather, examples of doctors' age-old habit of playing God, or what Haraway has called the "god-trick" (1991, 195), this time through the practice of writing, in which doctors appropriate not, or not only, their patients' bodies, but their patients' stories as well. Do doctors' narratives enact a resubjectification in speaking for their patients in literary language rather than in scientific language? Or, to paraphrase Gayatri Spivak's question posed to Deleuze and Foucault (among others, of course), "Can the subaltern patient speak?" In her important essay (1994), Spivak investigates the ways in which subalterns are represented in the work of leftist intellectuals in general and Deleuze and Foucault in particular in the conversation between them published in *Language, Counter-Memory, Practice* (Foucault 1977b). According to Spivak, leftist intellectuals have no difficulty representing "politically canny subalterns," and, at the same time, representing themselves as "transparent" (1994, 70). Spivak reveals the "epistemic violence" that takes place in such homogenizing representations, in which the "Other of Europe" is only rendered as the "Self's shadow," not as subject itself. In order to counter such homogenizing representations, Spivak maintains, "one must . . . insist that the colonized subaltern *subject* is irretrievably heterogeneous" (79; emphasis in original).

In *Fruits of Sorrow* (1997), Elizabeth Spelman asks questions related to Spivak's: "who gets to suffer?" and "whose suffering counts?" Why are certain persons allowed to be the subjects of suffering, and why don't others get to be? Or, to put it another way that relates to the primary focus of *Treatments:* who gets to tell illness narratives? Spelman recognizes that there is an economy of suffering, in which some people's suffering and their stories of suffering are deemed worth more than other's. Spelman's "rhetorics of suffering" seems particularly pertinent with regards to literary representations of AIDS, especially since the literature on AIDS has been written predominately by gay men in the West. I have found no autobiographical accounts of the experience of AIDS from a non-Western perspective, which may point to the (Western) limits of the memoir form.[33]

Perhaps the closest non-Western account of AIDS in memoir form that I know of is Jamaica Kincaid's *My Brother* (1997), which is also of particular interest with regards to how the rhetorics of suffering operate. Although she herself now lives in New York City, Kincaid's memoir, to some extent, deals with AIDS outside the West, as she describes her brother's illness and his eventual death from AIDS on the island of Antigua. From Kincaid's title, the reader is immediately led to believe that this is Kincaid's brother's story. She begins by telling us that, though she doesn't know her brother, she loves him nonetheless; or, at least, she wants to love him. It seems from the very beginning that she is, in fact, not very interested in writing about her brother, and even less interested in knowing him; but is mostly interested in writing about herself in relation to this entity called "my brother" (it is telling, I think, that her brother is not even named until well over halfway into the book). Is *My Brother* about Kincaid's brother's suffering, or is it about her own suffering in relation to her family and upbringing in Antigua in general and in relation to her powerful mother in particular? Kincaid's brother becomes, then, something of an alter ego, an-other side to her self, her "Self's shadow." She portrays him as irresponsible, sexually and otherwise; and she is both appalled and awed by what she perceives as his death drive. In the course of her return to Antigua and the family space she has willingly left behind, she comes to learn something crucial about her brother—that he was homosexual (unbeknownst to the entire family, who saw him as a "womanizer")—though she never comes to know him. She insists that she will never forget him; although this has nothing to do with who he was, and everything to do with who he was for Kincaid: that is, as the person who led the life she managed to escape from by leaving Antigua and her mother.

She explains that he lived his life in shadows, and in Kincaid's telling he remains in the shadows. Perhaps Kincaid senses that she can do little more than describe this shadowed quality to his life, rather than illuminate it. What burns bright in this book, however, is Kincaid's own fury, her resentment at her mother, who discouraged her from leading the life of a writer by burning her books as a child and by refusing to acknowledge her literary talent as an adult. Her resentment is at her mother, and by association at her brother, whose arrival when Kincaid was 13 years old changed her relationship to her mother, because it changed her family, making them poor and without hope, or so Kincaid believes. Kincaid wants to replace the burning books of her childhood with books that burn bright with the truth of her mother's failures. The reader is scalded by her words and by the truths she tells, that can only ever be her own. Kincaid admits she became a writer to save herself, and says that when she learned her brother was dying, her response was to write about him. But, has she in any sense written about him; or, has she appropriated his suffering as a vehicle for her own suffering? We will never have his story; her brother will forever be only parenthetical to Kincaid's own story.[34]

The same might be said of those works that attempt to describe the suffering of "other" populations affected by HIV/AIDS in the West, which tend to be written by doctors caring for PWAs in the inner cities (like Campo, as well as Zuger 1995 and Scannell 1999) and more rural areas (like Verghese) rather than by the PWAs themselves. In her attempt to delineate a rhetorics of suffering, Spelman negotiates the "use and abuse of suffering" employing the historical example of white American suffragists in the nineteenth century, who appropriated the experience of slavery as a metaphor for white, middle-class women's experience, and the constraints—moral, social, and legal—placed upon them that prevented free movement within the private and public spheres. As Spelman notes, many nineteenth-century suffragists had been and continued to be active in the abolitionist movement, and "drew heavily upon the language and imagery of the experience of slavery to make sense of and bring attention to the social, legal, and economic constraints under which they lived" (114; see also Davis 1983).

Although Spelman recognizes that to a certain extent such appropriations make sense, nonetheless, she asks that we consider the possibility that some sufferers are made into "spiritual bellhops," that is, "carriers of experience from which others can benefit" (8). In Spelman's analysis, then, slaves become "spiritual bellhops" for white, middle-class suffragists, who are "relieved that they [slaves] actually have had the experiences we [white suffragists] simply want to have the appearance of having had" (119). We might ask, then, if doctor-writers, like the white suffragists who appropriated the stories of slaves and like Kincaid, who appropriated her brother's story, are using their patients as "spiritual bellhops." Are they seeking their own redemption and compensation (both spiritual and financial) in the appropriation and commodification of their patients' stories?

Like Spelman, who is careful not to imply that experiences cannot be in many senses shared, and that the sharing of experience, moreover, is one of the purposes of tragic art, I do not want to deny the possibility of shared experience in the doctor-patient relationship. Nonetheless, I also want to consider another possibility, that, as Spelman puts it, "humans are ingenious in devising ways to deny such shared status even when appearing to affirm it" (132). With regards to doctor narratives, it isn't so much that these doctors simply appropriate their patients' stories as their own; rather, they take their patients' stories and craft them into works in which they claim that the true subjects of the stories are not the authors themselves, but their patients. And yet, they (the doctors) are the heroes of these stories, while their patients must bear the burden of suffering.

There are, it seems to me, at least three possible and interrelated compensatory fantasies operating in such narratives, which I call the documentary fantasy, the fantasy of identification, and the fantasy of empathy. Through these fantasies, doctors are able to give the appearance of becoming-patient and affirming a shared status between themselves and their patients, while at the same time appropriating their patients' stories, and in turn denying this shared status. Simply put, doctors are able to give the appearance of documenting their patients' stories, identifying with their lives, and empathizing with their experiences. They tell us they are documenting, identifying, and empathizing, but I question whether they are doing what they say they are doing. In what follows, then, I look briefly at the documentary fantasy in Abigail Zuger's *Strong Shadows* (1995) and the fantasy of identification in Verghese's *My Own Country*, before turning to a more extensive analysis of the fantasy of empathy in Campo's *Poetry of Healing*.

The Documentary Fantasy

In *Strong Shadows*, Abigail Zuger, a doctor working at an inner-city AIDS clinic in the late 1980s and early 1990s, attempts to offer a more complete portrait of HIV/AIDS in America. According to Zuger, the American public has been presented—"in novels and movies, sitcoms and confessionals"—with only "one face of AIDS," by which she means the face of a white, middle-class, gay male.[35] It is Zuger's hope to bring other faces of AIDS out of the shadows cast by urban poverty and drug abuse. "I wrote this book to fill in the faces," she maintains, "to give them mouths and let them speak. Our patients are far more eloquent historians of their times and their disease than media reporters and salaried spokespersons alike. I didn't want their part of the history of AIDS to be forgotten" (1995, xi).

Zuger believes that the ideal means in which to do so would have been to mount a camera in her examining room, "point it at the chair next to my desk where the patient sits, and let it run" (xii). She seems to think that this method would have most *realistically* captured her patients, by focusing exclusively, and in close-up, on them. Notice, of course, that the doctor is out of the frame—literally and figuratively—in Zuger's ideal scenario. Despite her "reality TV" sensibility, however, Zuger must settle for another method of recording her patients' stories.[36] "Failing this option I have tried to achieve the same result with words," Zuger explains. "I have changed names and some details to protect the identities of patients and their families, but otherwise, *this is a documentary book*" (xii; my emphasis). Zuger insists that she has kept herself and her co-workers at the clinic in the background in order that "[their] patients have the stage" (xiv), but this is little more than fantasy, Haraway's "god-trick," in which Zuger as author positions herself above the situations she writes about.

While Zuger does succeed in detailing the ways in which "AIDS in the inner city has less to do with sexual politics than with individual citizenship, less with abstractions of art and mortality than with the mundanities of food, shelter, child care, and Medicaid cards" (xii), nonetheless, she does not succeed in keeping herself out of the picture. In fact, although each chapter bears the name of one of her patients as if they are in the chair in front of the camera telling their stories, there is never any doubt about who is running this show, who is the authority on suffering, who is authorized to tell of suffering, and who, indeed, is the author of these tragicomic vignettes of AIDS in the inner city. It isn't so much the fact of Zuger's representation that is problematic; she offers a compelling portrait of a doctor who not only listens with her stethoscope and other instruments to her patients' bodies, but also removes her stethoscope and takes a seat across from her patients in order that she may better hear their stories. What is troubling is the manner of her representation; in particular, it is her attempt to portray her work as something it isn't and could never be: a documentary of her patients' lives unmediated by her role as medical and literary authority. Zuger claims she is out of the picture, that she is transparent, just as Spivak's leftist intellectuals claim to be, even as the narrative is driven inevitably by her subjectivity. The emphasis, therefore, is on what she can and cannot do for her patients, as well as on her interpretations of them as patient character types, including, for example, polite and easy to deal with (Michael Soto), confrontational and difficult (Deborah Sweet), withholding and difficult (Anita Lewis), or out to cheat the system either for financial gain or to meet a pathological emotional need (Shannon Gallagher). Zuger, then, is like Spivak's "first-world intellectual masquerading as the absent nonrepresenter who lets the oppressed speak for themselves" (1994, 87).

The Fantasy of Identification

Like Zuger, Verghese wants to present stories of AIDS that, before his book, were generally unknown. Unlike Zuger, however, he doesn't fantasize that his rendering of his patients' lives will be documentary; rather, he likens the names at the top of each blank chart to "the title of a novel that you had just bought, the jacket cover still pristine, the book new" (92). He doesn't, therefore, deny the literariness of these stories, and this is reflected in the representation of the return of the first AIDS patient to Johnson City with which he opens his memoir. Meeting a new patient for Verghese is stepping across a threshold into "a limbus of time, a labium of space" (91); what awaits him might be the first paragraph of an "apocalyptic short story" that grabs the reader so that he or she can't put it down (92). The subtitle to *My Own Country* is revealing: *A Doctor's Story of a Town and Its People in the Age of AIDS*. Like Zuger, he (and/or his publishers) implies that the focus is not on him. Never mind that the subtitle's signaling of an ethnographic approach contradicts the individual, autonomous subject signaled by "my own country." Nonetheless, in *My Own Country*, Verghese confronts not only assumptions about who contracts HIV, but also class biases that prevent certain stories from being told. Although Spelman's questions about whose suffering counts are not specifically directed at the publication and reception of memoirs, at the turn of the twenty-first century memoirs have become a popular form in which many people attempt "to make sense of suffering" (Spelman 1997, 4), either by writing them or by reading them. But, again, the question is, whose stories are really being written and read?

As I noted in my discussion of Verghese's work in the first half of this chapter, he is driven by a desire for home, and a desire to belong. He knows he is an outsider in Johnson City, Tennessee, and in the institution of American medicine, but it is precisely this status as outsider that allows him to identify with, or so he believes, the gay men and others who he meets in his AIDS practice and who are stigmatized because of their illness and its link to "illicit" behaviors. He sees parallels between the experience of "resident aliens" and gay men in America; both, according to Verghese, must become "experts at blending in, camouflaging themselves." But, for Verghese, this camouflaging of self is "voluntary, even joyful," rather than, as he thinks it must be for gay men, "a great cost to their spirit" (51). On the one hand, then, Verghese identifies with what he understands as the necessity of faking conformity, but, on the other hand, he characterizes his own faking conformity as a happy habit of reinventing himself, whereas he presumes the gay man's faking must arise out of shame and guilt.

Verghese is ever curious about his gay male patients' lives down to the details of their sexual practices. He notes that while he "took delight" in these details, his transcriptionist was scandalized: "'Campy,' 'queen,' 'fisting,' 'rimming,' and 'water sports,'" he explains with a knowing wink, "were not in the medical dictionary that sat next to her keyboard" (103). Although he identifies with—indeed, takes delight in—their outsiderness, at the same time he exoticizes gay men and their sexual practices, and frequently pathologizes them. Thus, a visit to the Connection, Johnson City's lone gay bar, to encourage men to get tested and practice safe sex becomes a risky voyage into a seedy heart of sexual darkness. Back at home and safe, he pours himself a stiff Scotch and admits, "I felt as if I had returned from a dangerous mission and had miraculously emerged unscathed" (59). Gay men are dangerous, as is AIDS care, and Verghese is drawn to such danger.[37] But he prefers his danger contained in the form of a fantasy literary identification, like his infection dream, rather than the real thing.

Toward the end of his time in Johnson City, Verghese meets Will and Bess Johnson, an elderly, long-married couple with grown children, much respected in their community, who have both been infected with HIV after Will received infected blood during heart surgery at Duke University Medical Center. Verghese is surprised to discover himself treating the Johnsons differently than his other patients—escorting them to their room, carrying their luggage, spending extra time visiting them—and he wonders if this is a "blind spot, a class prejudice" on his part. Not only are the Johnsons not outsiders to society, they are also widely regarded as having contracted HIV through supposedly "innocent" means. Nonetheless, they are unwilling to be open—even to their own children—about their illnesses, and Verghese, who is more than a bit dismayed by their silence, does not voice, to them at least, his belief that their social status might counter some of the "stigmatizing explanatory narratives" attached to HIV/AIDS (Epstein 1995, 169).

Ever self-reflexive, Verghese attempts to explain this "blind spot" made apparent by his association with the Johnsons in Freudian psychoanalytic terms. He thinks his "egalitarianism"—his nonjudgmental dealings with gays and other social outcasts—might be a reaction formation to cover up "a deeply rooted prejudice." But, what if his "blind spot"—his class prejudice—is a reaction formation itself: against the possibility and threat of dangerous identifications? Indeed, what does his identification with the Johnsons cover up? His prejudices or *other* identifications that might reveal his dangerous desires? In other words, while his identification with the Johnsons and his literary representation of his encounters with gay men and their sexual practices might be a reaction-formation against a deeply rooted prejudice, it might also be a reaction formation against a deeply rooted desire. I don't necessarily want to resolve what is clearly a conflict—between safety and danger, between belonging and not belonging—at the core of Verghese's presentation of self, rather I want to point out the ways in which the safe confines of literary narrative allow him to identify with his patients *and* keep them at a distance. Zuger's fantasy is that she is transparent within the stories she tells; Verghese's fantasy is that he can get close enough to know his patients' stories, but not so close that their dangerous desires become his own. Campo's fantasy, as we shall see, is that his patients' stories are his own, that they are, in fact, *about him*.

The Fantasy of Empathy

In order to investigate Campo's education in empathy, I want to focus on a particular chapter in *The Poetry of Healing* that I didn't take up at all in my earlier discussion of his work. "Fifteen Minutes after Gary Died" tells of Campo's friendship with Gary Fisher, an African-American writer with AIDS, who he meets through his former writing teacher at Amherst—the person who inspired him to write—Eve Kosofsky Sedgwick. Fisher was also Sedgwick's student, and Campo's initial response to him seems to be curiosity spiced with something like jealousy over the fact that he is not the teacher's only pet.

After their initial introduction, Campo doesn't meet Fisher again until he is admitted with complications from AIDS to the ICU at the University of California at San Francisco Hospital, where Campo is a resident. When Sedgwick explains that Fisher is HIV+, Campo admits, "now for the first time I actually and personally knew someone with AIDS" (1997, 133). Considering that Campo was living in San Francisco in the early 1990s and gay himself, this is rather stunning news. But not nearly so stunning as the fact that he then earnestly notes, "In my selfish experience of that moment at his bedside, I felt the shock as though I myself had received the terminal diagnosis. No longer could I consider myself immune or invisible; it was as if my body's defenses were suddenly as pulverized as Gary's, and my own decrepit body lay outstretched before me" (133). Because they share certain characteristics—they are both gay, nonwhite, writers whose mentor is Eve Sedgwick[38]—Campo recognizes himself in Fisher, and this recognition allows him to feel empathy. Fisher, Campo believes, is his double, and, thus, his (Fisher's) diagnosis is his (Campo's). This recognition—this empathy—makes Campo certain that he *knows* the shock of receiving a diagnosis of death. And yet, of course, he has not received a diagnosis of death at all. His body's defenses are not pulverized, nor is he decrepit, lying outstretched on a hospital bed.

Through his relationship with Fisher, who is never actually his patient, Campo becomes-patient in the manner that I described earlier. But what is the limit to such a practice? At what point does empathy become appropriation? This limit is what Lacan attempts to delineate in his ethics seminar;[39] it is the limit of language in describing traumatic experience, such as pain, suffering, and death, and also the limit of the body itself. Although the body is never inviolate, nonetheless, we cannot know or share another person's experience of embodiment. Illness narratives are an attempt to communicate an experience of embodiment, but they cannot communicate it exactly: words cannot give us the thing itself. Campo seems completely unaware of the possibility that his knowledge and, indeed, his language are finite. He does sense that his "fantasized illness" is self-indulgent, but doesn't sense that such self-indulgence might rely on using another person as a "spiritual bellhop":

> I began to be infected with AIDS myself, I imagined, to need the virus in the same physical way as I came to need Gary's voice. I wallowed in my own fantasized illness with Gary's help, for the first time in months allowing myself to identify not dangerously but joyously with someone who was suffering, and thus thwarting the effect some of my residency evaluators had been having on me. I was completely self-indulgent, cultivating and tending to the KS growing in my lungs and on my skin, feeling the pitter-patter of the tiniest pneumocystis organisms in my airways, the insidious suffocating feeling prompting me to cry out with whatever happiness and pain I discovered in my heart, the new purple blotches on my arms an embodied refusal to remain invisible. (139)

If the question for Verghese is what is the danger in his distancing literary identifications, his joyous faking conformity, the question for Campo is what is the danger in his displacing literary identifications, his joyous faking suffering?

Campo reads forgiveness at the heart of Fisher's writing, and at the heart of his very being; and yet, forgiveness is also precisely what Campo needs to read in Fisher's writing. Is forgiveness, then, what Fisher's writing enacts, or is it what Campo wants and needs it to enact? Campo admits that for Fisher "the opportunities for rage were abundant, and certainly rage and even despair found ample expression in the work." Nonetheless, he also insists that, "ultimately it was Gary's utter decency and sweetness that burrowed its way into my own heart" (148). Campo fits Fisher's writing neatly into what I have called the American model; forgiveness, finally, must overcome rage and despair. But, it is ironic that Fisher's writing should be precisely what the doctor—Campo—ordered. Gary's writing—and, indeed, "Gary's spirit"—becomes meaningful only in terms of its meanings for Campo: it is through the "availability of [Gary's] suffering" that Campo is able to bring himself closer to the suffering of his patients. Gary's suffering and the "generosity of Gary's spirit" make Campo guiltless: he is neither a victim nor an agent of affliction. Fisher's affliction, in other words, allows Campo to remain unafflicted.

Not only does Campo make use of Fisher's suffering, but he also transforms his suffering into an attempt on Fisher's part to get Campo's attention. When Fisher ends up in the ICU again, unable to breathe, Campo reads Fisher's inability to breathe and his call to the ICU as a call to Campo himself, simply because, if he had arrived at the hospital just a few hours later, he would have been Campo's patient. "By the sheer force of his illness," Campo intones, "as if it were somehow aware of what Gary wanted, as if Gary even at this late stage of the disease still managed to control and harness a bit of its destructive power, the relationship of doctor to patient was nearly imposed upon me" (153). And, according to Campo's narcissistic reading of his suffering, what Fisher desires, most of all, is Campo's attention, for Campo to be his doctor. Yet, despite the amazing capacity of his illness to reveal Fisher's intentions, Campo also tells us, "He was in the hospital for several days before I even knew he was there" (153). Campo creates an intimate relationship where there appears to have been none.

When Campo does finally visit Fisher, he tells us that Fisher was "pissed off," presumably with Campo, and "in pain." But, despite his anger and pain, we are told he once again forgives Campo, and sends him "out into the world as an embodiment of what he thought might be possible to change in that world, that alien place AIDS had created for him, and now for me, to decipher" (155). Once again, it is Fisher's suffering that allows Campo to move out into the world, potentially to decipher and transform it. While Fisher's world flattens and contracts, and even taking a breath requires assistance, Campo's world expands, and his powers—to move, to decipher, and to transform—increase.

In the doctors' narratives that I have looked at in this chapter, is there an occultation of patients' lives? Are these representations in effect doctors' self-portraits in which the subjectivity of the patient is effaced? Is Verghese's position voyeuristic in relation to his patients' stories, does he maintain a safe distance? Is his a joyous faking not of conformity, but of nonconformity? And, is Campo's position fetishistic in relation to his patients' stories, and in particular Gary Fisher's story? Is his representation belated because he missed Fisher's death or because he missed Fisher's life?[40] In Campo's representation of Fisher's death, there is no death and no Gary; rather, Gary simply fades away. The chapter ends not with a portrait of Fisher's dying, but with a self-portrait of Campo as survivor. With survival comes the suffering of guilt, but Campo's survival and his guilt is always already forgiven. Forgiveness is always right on time in this scene of one man's survival and another's fading. But, paradoxically, Campo also fails to be the hero of this story; he fails to offer his patients a cure for AIDS, and he often fails to empathize with them; he fails to be Gary's doctor and his friend; and he fails to write of what it is like to be dying. And yet, forgiveness may be all any of us has to offer—to ourselves and each other—because there is, ultimately, no safe position in the world. We must focus, then, not only on the representations themselves, but on the limit of representation in general; it is this limit, this failure to represent the whole story, or even enough of the story, that destabilizes our false sense of security, and points towards other possible renderings.

I want to conclude with an acknowledgement of the ways in which this chapter has inevitably effected some of the violence it meant simply to illuminate. In particular, since I have focused on doctors' narratives, I have thus far failed to mention (except obliquely in a footnote) that some of Gary Fisher's own writings—on his experiences of both sexuality and dying—are available in print. In *Gary in Your Pocket: Stories and Notebooks of Gary Fisher* (1996), edited and published posthumously by Eve Sedgwick,[41] the reader is presented with Fisher's desire for "the outrageous, the unorthodox and unacceptable," first in sex and, later, in treating his illness. Fisher recognizes the inextricable link between his sexuality and his disease, but not in the usual sense, which would blame the disease on his sexuality. Rather, Fisher understands the outrageousness and unorthodoxy that characterizes his sexuality not as the cause of his disease, but as a resource to be utilized in his treatment. Fisher knows that his illness, like Flannery O'Connor's lupus, might not only cripple him, but also, potentially, drive him "like a fast car" (258–59). His illness drives him to write "too soon" (261), to close the gap between his experience and his words. Does he forgive his caretakers as Campo believes he has forgiven him? His impressions of the doctors, nurses, and technicians are nothing if not ambivalent: their multiple invasions of his body make him "wonder what they don't know" rather than what they do know. "Seems I know they're trying to help me," he writes in his journal, "but the snowball effect tends to be cold and painfully circular. [. . .] And to this point I've seen nothing but" (261). The odd locution of "Seems I know they're trying to help" reveals his fundamental ambivalence: the "seems" negates the "I know they're trying to help."

Unlike Campo, Sedgwick is aware of her formative role in this presentation of Fisher's writing to the public. In particular, she acknowledges that, although she and Gary agreed upon the title before his death, it risks "trivializing" Fisher's work. Sedgwick explains her predicament:

> In metaphor, it seems to make Gary small, appropriable; and it may be all too resonant in the context of the posthumous publication of an African American writer, mediated by an older Euro-American editor and friend, from the press of a mostly white, Southern university. Gary and I were both very conscious of a history of white patronage and patronization of African American writers, the tonalities of which neither of us had any wish to reproduce. Sexuality was a place where Gary was interested in dramatizing the historical violences and expropriations of racism; friendship, authorship, and publication, by contrast, were not. (285–86)

Nonetheless, Sedgwick goes through with her project to bring some of Fisher's writing into print. She knows *Gary in Your Pocket* (both title and edited work) is inadequate; she also knows it is necessary not only to satisfy "the survivors' yearning for a potent, condensed, sometimes cryptic form of access to the person who would otherwise be lost," but also to allow the dead a space "to continue to resist, differ, and turn away from the living" (286). It is this space for the dead to turn away from the living that Campo fills with the forgiveness *he* requires.

5.

Between Two Deaths

Practices of Witnessing

I have discussed many of the ways that illness narratives read and write the body, but I have also sought to explore the ways that they demonstrate the impossibility of reading and writing the body in any complete form. Illness narratives bring up a number of questions about the relationship between language and embodiment: Is the experience of embodiment determined and structured by and in language? What is lost in the attempt—the urgency, even—to bring the body to language? Can we encounter the body outside of or prior to language? Can we tell stories and bear witness, not only *about* the body but also *through the body?*

In this chapter, I look at the constitutive losses that lead to new forms of telling and listening in Paul Monette's autobiographical writings on HIV/AIDS, in which he mourns the loss of his partner, Roger Horwitz, and in John Bayley's trilogy[1] on the demise of his wife, Iris Murdoch, to Alzheimer's disease. Monette's and Bayley's works might be thought of as, in Alice Kaplan's terminology, "language memoirs" (1994, 59). Although Kaplan's own work investigates the relationship between language and identity in the process of learning French and becoming a teacher of French, her conception of language memoir is more broadly conceived than her own language memoir would suggest: "the second language is not always a 'foreign' language; sometimes it is a new dialect, a language of upward mobility, a language of power or expressivity within the native language" (59). In Kaplan's formulation, the second language can be either a private or public language, a language of dream and escape, or an expert and instrumental language. In Monette's and Bayley's cases, the experience and event of illness often requires that new languages be learned or even invented.

The question of appropriation, which I problematized in chapter 4, is of concern in this chapter as well. Although it will not be the main focus of this chapter, nonetheless I think it is important to consider the possibility that one half of a couple might appropriate the other half's illness and story about that illness for personal gain. As with the relationship between doctor and patient, the relationship between caregiver and the person cared for is often hierarchical and dichotomous. Bayley, in particular, has recently come under fire in the British press for exploiting the memory of Murdoch for his own gain. Although his first book was greeted, even in the skeptical British press, as a loving portrayal of his remarkable marriage to Murdoch, the second and third volumes were received far less kindly by critics in particular and the reading public in general. It should be kept in mind, however, that the works of Monette and Bayley do not pretend to be collaborative efforts, as, say, *Cancer in Two Voices* and, even, *Before I Say Goodbye* claim to be.[2] Rather, what they foreground is the interminable process of mourning that the illness of one's partner—and the many losses that it brings—introduces into a relationship that had been, prior to the illness, more or less taken for granted.

The experience and event of illness initiates a crisis of embodiment as well as a crisis of language. Or, perhaps, a crisis of embodiment is always also a crisis of language. The experience of illness requires a new attention to the embodied self as it negotiates the world, and a new attention to language. Psychoanalysis, especially as formulated by and after Lacan, is concerned with the ways in which language is negotiated, particularly in situations of crisis or trauma. Psychoanalysis is both a theory and a practice; or, more to the point, it is a theory that emerges out of what I call practices of witnessing.[3] Intersubjectivity in general and specific relations in particular—between doctor and patient, between a person who is cared for and a person who is doing the caring, between a person who dies and a person who survives—are at the forefront of my analyses. Affective history, as I understand it in and through illness narratives, is concerned with how experiences of illness are brought into language, and what the difficulties are in doing so.

To do affective history is to negotiate with what Lacan called the "real." The Lacanian real is that which exceeds—is beyond the limit of—signification and cannot be brought into language. Although it absolutely resists symbolization, at the same time, paradoxically, the real is the kernel around which all symbolization is structured. The real is like the number zero; it is that which isn't a number, and represents nothing (no-thing, no object, no subject), but is also that which begins the sequence of numbers or the signifying chain. The real stands as the midpoint between positive infinity and negative infinity; it is the empty space between measureless loss, everything that has been, and measureless possibility, everything that might be. In *The Four Fundamental Concepts of Psychoanalysis*, Lacan discusses the real in relation to trauma, which he defines as that which is "unassimilable" into conscious experience (1981, 55). Lacan describes the real in spatial and temporal terms; it is the place of an encounter that is, paradoxically, always missed because we are always too soon or too late for it. The real insists in "making us aware of its existence," but we never see "its very face, [we see only] the screen that shows us that it is still there behind" (55).

What does the real have to do with the experience of illness and writing about that experience? As many of the illness narratives I discuss show, illness is a traumatic experience that is often apprehended as a rupture. This experience of rupture, as Lacan conceives it, is a "nontemporal locus . . . which forces us to posit . . . the idea of another locality, another space, another scene, *the between perception and consciousness*" (56; emphasis in original). Many illness narratives attempt to describe this other locality, and they usually do so in spatial terms, emphasizing that illness is often experienced as being in a strange place, or, as Monette describes it, "on the moon." Yet, they also attempt to describe the ways in which illness disrupts time and creates another temporality. The narrative itself and the practice of writing may give form to the experience of illness, but it is often a form that confounds a linear structure with a clearly marked beginning, middle, and end.

In his seminar on ethics in general, and with regards to his reading of the figure and story of Antigone in particular, Lacan speaks of the ethical possibility of being *between two deaths*, which in his formulation is an ethics that emerges both from a different spatiality and a different temporality. Antigone opposes the laws of the city—which are, of course, also the laws of her uncle, Creon—when she attempts to bury her brother, Polynices, in opposition to Creon: in opposition, that is, to the law of the father. By her action, Antigone exiles herself from the law—the law of the father, the symbolic order—and as punishment, is buried alive in a tomb. Her transgression places her in an impossible position, which, according to Lacan, is the position of being *between two deaths*. In an explication of this Lacanian concept, Zizek writes that, "Lacan conceives this difference between two deaths as the difference between real (biological) death and its symbolization"; and, "in Antigone's case, her symbolic death, her exclusion from the symbolic community of the city, precedes her actual death and imbues her character with sublime beauty" (1989, 135). Antigone is neither of the living nor of the dead; she is in between, and it is this inbetweenness that language cannot describe, because it is outside of and beyond symbolization.

In their practices of witnessing to the loss of their beloveds, Monette and Bayley attempt to articulate this inbetweenness, this gap between a first failed death and a second absolute death. For Monette, the first death is the death of his lover; Roger's death is the death of an other, but because, in a sense, Monette will die the "very same" death, it is both the death of the other and the death of the self that he must witness. There is an echo of Monette's own death that reverberates beyond Roger's death and throughout Monette's writing.

In the case of Bayley's narrative of Murdoch, the first death is the death of the self through Alzheimer's that precedes the death of the biological body. Alzheimer's is a tragic disease in every case, but in Murdoch's case it is especially tragic, or, perhaps, more accurately, *ironically tragic*. As a philosopher and novelist, Murdoch formulated a theory of the self that, paradoxically, called for the dissolution of the self. She saw tragic art as providing the means by which one might undermine the narcissistic self, which makes Alzheimer's—what Paul John Eakin has called "the death of the self" (1999, 46)—a particularly ironic double death for Murdoch. Bayley attempts to bear witness to Murdoch's ironic double death, to communicate with her across the gap that separates her from him.

Monette's and Bayley's works show the ways in which witnessing is not comprised of a single act of telling or listening, writing or reading, but is a practice that must be maintained even, or especially, as it appears to fail. In his essay on AIDS diaries and the literal and figurative death of the author, entitled *Facing It* (1998), Ross Chambers describes witnessing as a mediating practice of mourning. Monette and Bayley reveal the means by which they act as relays, as Chambers puts it, to carry the message deferred by death forward (26). Being between two deaths puts one in a unique position to carry the message forward, and Monette and Bayley take seriously their responsibility to act as relays from beyond the limit and back again.

Monette's Vigilant Witnessing to AIDS

On October 22, 1986, Paul Monette's lover Roger Horwitz died of AIDS. "That is the only real date any more," Monette writes, "casting its ice shadow over all the secular holidays lovers mark their calendars by" (1988a, 2). In the year following Roger's death, Monette, himself HIV+ and up until then known (or not known as the case may be) as a writer of rather banal novels, earnest poetry, and film novelizations, wrote two works—*Love Alone: Eighteen Elegies for Rog* (1988b) and *Borrowed Time: An AIDS Memoir* (1988a)—that bear both personal and public witness to the early days of the AIDS epidemic in the United States. Before his own death of AIDS on February 10, 1995, Monette would write two other autobiographical works: *Becoming a Man* (1992), which describes his torturous coming out, and *Last Watch of the Night* (1994), a collection of "essays too personal and otherwise," that chronicle his continued practices of witnessing to AIDS. It is no small irony, then, that AIDS would both give Monette his voice and also mortally wound him; his voice and his wound are inextricably bound to each other "as the condition of the possibility of telling" his story and the story of others (Jacobs 1994, 3).

The voice that emerges in Monette's writings on AIDS is an ethical voice; it is a voice of witness connected intimately to his experiences of loss, love, and mortality (his own and others). "I buy time with another story," Audre Lorde writes (1997b, 409), but Monette understands that time cannot be bought but merely borrowed, implying a debt that casts its shadow on the future. And so, *Borrowed Time* opens with the jarring statement, "I don't know if I will live to finish this." Monette continues:

> Doubtless there's a streak of self-importance in such an assertion, but who's counting? Maybe it's just that I've watched too many sicken in a month and die by Christmas, so that a fatal sort of realism comforts me more than magic. All I know is this: the virus ticks in me. And it doesn't care a whit about our categories—when is full-blown, what's AIDS-related, what is just sick and tired? No one has solved the puzzle of its timing. I take my drug from Tijuana twice a day. The very friends who tell me how vigorous I look, how well I seem, are the first to assure me of the imminent breakthrough. What they don't seem to understand is, I used up all my optimism keeping my friend alive. Now that he's gone, the cup of my own health is neither half full nor half empty. Just half. (1988a, 1–2)

Monette did live to finish the book, but is there something more, beyond the book, that he senses he will not live to finish, that cannot be finished in language, that cannot be fully told? When he writes that, "the cup of my own health is neither half full nor half empty. Just half," Monette attempts to explain his predicament, which is the predicament of the witness, of the survivor. The "just half," in Monette's formulation, resists symbolization and defies any easy interpretation that would reduce this to a story of hope ("half full") or a story of hopelessness ("half empty").

Like *Borrowed Time*, *Love Alone* also opens with a "point of failure of symbolization": "everything extraneous has burned away," Monette writes in the poem "Here" (1988b, 3). And, what follows in both books is not a record of a traumatic event as it really occurred in its positivity, but rather the traumatic event "constructed only backwards, from its structural effects" (Zizek 1989, 169; see also Lacan 1977). Monette's work is about the absolute necessity of telling of death (the death of Roger in the past, the death of Paul in the future, and the epidemic of deaths from AIDS in the past, present, and future) as well as the impossibility of comprehending the meaning of this death. In order to show the enormity of both a single death and countless deaths, Monette returns again and again to personal and political scenes of loss, but also, importantly, to personal and political scenes of love. He explores different genres—memoir, poetry, essay, and fable—in order to find a suitable form.[4] What he discovers in this exploration of form, and the reader discovers in reading his work, however, is that there is no form particularly suited to what he urgently needs to say. Rather, his work achieves its emotional poignancy through the conflict between the urgency of Monette's witnessing and the inadequacy of the literary form to convey his message.

In the influential early collection of AIDS criticism, *Writing AIDS* (Murphy and Poirier 1993), several authors read Monette's work as exemplary of particular forms of AIDS writing. John M. Clum calls Monette the "paradigmatic writer in this new barren land of displacement, pain, and loss," and "the bard of AIDS" (Clum 1993, 209–10). In his comparison between two modes of AIDS writing, which he calls immersive and counterimmersive, Joseph Cady reads Monette as a classic example of the immersive mode. According to Cady, immersive writing attempts to confront the denial surrounding AIDS by thrusting the reader "into a direct imaginative confrontation with the special horrors of AIDS" (Cady 1993, 244). Counterimmersive writing, on the other hand, portrays AIDS tangentially. Cady favors the immersive form because he fears that counterimmersive writing "runs the risk of ultimately collaborating with the larger cultural denial of the disease" (261). Finally, Timothy F. Murphy's essay, "Testimony," provides an important discussion of AIDS writers, including Monette, who write about the epidemic in the testimonial form. Murphy defines testimony as "witness in front of an indifferent world about the worth and merit of persons. And thus one writes, for the world unconvinced, that someone was here and that, death notwithstanding, a presence remains" (Murphy 1993, 317).

My analysis seeks to build on the AIDS criticism of Murphy and others by reading Monette in relation to theories of witnessing that have developed roughly concurrently with the AIDS crisis. While I am aware that there is a large body of literary critical work on writings about AIDS in general and Monette's work in particular, my approach diverges somewhat from a conventional literary critical reading of Monette's work toward a more phenomenological examination of its philosophical grounding and cultural context. As in the previous case studies presented here, I seek not only to read such literature through contemporary theories of subjectivity and the body but also to read those theories through the event and experience of various illnesses, including AIDS, as described in literature. Thus I read Monette's work in particular along with and through the work of the contemporary feminist philosopher Kelly Oliver. All of Oliver's work asks questions about the relationship between subjectivity and ethics, and in doing so, engages with a wide range of twentieth-century continental philosophers, especially those influenced by a psychoanalytic framework, including Lacan, Jacques Derrida, Luce Irigaray, and Julia Kristeva.

In *Witnessing: Beyond Recognition* (2001), Oliver articulates a theory of witnessing that draws on and extends recent philosophical work on recognition as the basis for subjectivity, as well as psychoanalytic-influenced trauma theory, especially as articulated by literary theorist Shoshana Felman and psychoanalysist Dori Laub in their influential book *Testimony: Crisis of Witnessing in Literature, Psychoanalysis, and History* (1992) and historian Dominick LaCapra in *Representing the Holocaust: History, Theory, Trauma* (1994). Trauma theorists, like Felman, Laub, LaCapra, Cathy Caruth (1995, 1996), Lawrence Langer (1991), and Maurice Blanchot (1995), have attempted to understand the relationship between trauma and experience, and the ways that certain traumatic experiences are given voice in written and oral testimony. Much of this theory has emerged out of an analysis of experiences of extreme violence, such as occurred during the Holocaust and other wartime events, but less has been written about illness as a traumatic event that, like violence, may be both necessary and difficult to witness. In her edited collection, *Trauma: Explorations in Memory* (1995), Caruth includes an interview with AIDS activists and cultural critics Gregg Bordowitz, Douglas Crimp, and Laura Pinsky, in which she asks them to discuss the ways in which the AIDS crisis can be viewed as trauma, which she defines as "a memory that one cannot integrate into one's own experience, and as a catastrophic knowledge that one cannot communicate to others" (256). I want to pose a question similar to Caruth's and explore the answer through Monette's writings on AIDS and Oliver's theory of witnessing. How does Monette's work—in both its content and its multiple forms—demonstrate what Oliver calls the "paradox of the eyewitness," which, as she describes it, is the "paradox between the necessity and impossibility of testimony"? (2001, 86).

Oliver notes that the word "witnessing" has a double meaning: it means both "*eyewitness* testimony based on first-hand knowledge . . . and *bearing witness* [to others] to something beyond recognition that can't be seen" (16). The practice of witnessing, then, requires that we cultivate our "response-ability," in Oliver's terminology, to those things that we both see and do not see, hear and do not hear, and know and do not know. For Oliver, "[t]o serve subjectivity, and therefore humanity, we must be vigilant in our attempts to continually open and reopen the possibility of response" (19). This openness to the possibility of response is a means by which we might, as Monette attempts to do, tell both personal and political stories of loss and love that surround the experience of AIDS. In her philosophical investigations into the practices of witnessing, Oliver is concerned with the possibilities engendered in "working-through the trauma of oppression necessary to personal and political transformation" (85).

How does one become a responsible witness in Oliver's terms, and what kinds of personal and political transformations are enacted by this sort of witnessing? We, the readers of Monette's work, are also implicated in this process of witnessing; we too must cultivate our response-ability through our reading (at the very least) to those whom our society, in the age of AIDS, has made "other." Where AIDS is concerned, bodies have been devalued and "abjected" (reduced to vectors of disease and dehumanized) not just because of the way in which they are perceived as particular sexual bodies, but also because of the way in which they are perceived as particular racial, classed, and national bodies. In his work, however, Monette is primarily concerned with the abjection of the gay body and love and/or sexuality between men.[5] My discussion of Monette is organized around three key terms in Oliver's work—history, vigilance, and working-through—and delineates her use of these terms and their place within an ethics and aesthetics of witnessing as performed in Monette's work on AIDS.

History

Drawing on the psychoanalytic work of Felman and Laub on Holocaust testimony, Oliver is concerned less with the historical accuracy of testimony than with the possibility that the "performance of testimony says more than the witness knows" (86).[6] Such is the case in Monette's testimonies to the AIDS crisis, and we will also see this phenomenon in Bayley's testimonies to the crisis that Alzheimer's creates in his relationship with Murdoch. Monette's AIDS writing not only entails what he knows about this illness and the process of learning about it and the death it brings, but it also reveals everything that he doesn't know, can't know, and even refuses to know. Monette's work performs what Laub describes as the "*discovery* of knowledge—its evolution, and its very happening" (1992, 62). His opening sentence in *Borrowed Time*—"I don't know if I will live to finish this"—is only the beginning of a chronicle of knowing and not knowing, of certainty and uncertainty, of recognition and lack of recognition. The crisis of AIDS is, in other words, an epistemological crisis as well as an ontological crisis. Monette's writing is an attempt to describe the impossible position of having both too little knowledge (to prevent or treat this disease) and too much knowledge (of the fact of death: Roger's, his own, and, in the beginning at least, seemingly everyone who is infected).

One reason perhaps Monette doesn't know if he will live to "finish this," is that he doesn't even know where to begin. Although he knows that, at the time of his writing *Borrowed Time*, it is the "seventh year of the calamity," he doesn't know when and where it all began (2). His opening chapter attempts—and fails—to pinpoint when he, and everyone else in the gay community in Los Angeles, began to know something. There are signs: a note in his diary in December 1981 about "ambiguous reports of a 'gay cancer,'" but, Monette admits, at that time, "I know I didn't have the slightest picture of the thing" (3). What is this thing, this anamorphic thing, as Lacan might say, that is imperceptible (and, for us, unreadable), even from a position seven years into it? How can we begin to look at this thing, begin to see it, begin to read it, begin to know it? As Oliver notes with regard to victims of oppression in general, but which might be applied to the experience of people with AIDS in particular, what one must seek is not merely "visibility and recognition," one must also witness "to horrors beyond recognition" (2001, 8). What would a history of this witnessing look, sound, and feel like? In his writing, Monette attempts to give this history, which is a history of practices of witnessing as much as a history of AIDS among gay men in the United States.

The difficulty for Monette, of course, is that he is *in it:* in the thing, in the calamity, not outside of it, or past it. Oliver asserts, again following Felman, that "it is impossible to testify from inside" a traumatic event, because the trauma possesses us so entirely that there is no outside of the event to which we might relate our experience.[7] And yet, Oliver continues, "in order to reestablish subjectivity and in order to demand justice, it is necessary to bear witness to the inarticulate experience of the inside" (90). Bearing witness to the inarticulate experience of the inside of AIDS is Monette's task, and it is an infinite task. His work attempts to give form to this inside that lacks parameters either in time or space.[8] When Monette first read reports of a "gay cancer" in 1981, he thought at the time: "How is this not me?" (3). This is a strange question; it is, in fact, grammatically strange, but more importantly it reveals *a grammar of estrangement*. That is, it reveals a knowledge of the self that is founded upon a failure of knowledge of the self. "How is this not me?" is a question, moreover, that structures all of Monette's writings on AIDS, and most narratives that attempt to describe the experience of illness or trauma from the inside. It is also a question that doesn't close off but, rather, opens up the possibility of response. And it is a question that reveals how one experiences AIDS in particular or illness in general, and also suggests a theory of subjectivity that resonates with Oliver's.

Other theories of subjectivity place the annihilation of difference at their center, and they understand identity as a fixed and stable category of being. Oliver, however, maintains that differences—how is this not me?—are what motivate her to "try to move out of myself towards you in order to commune with that which ultimately I can never know" (1997, 96). She continues, "it is through our relationship and our *differences* that I can begin to see something of myself. . . . We experience our lives as flux and flow, full of surprises even to ourselves" (96–97). Monette's question is significant simply because it is a question. When he asks, "How is this not me?" he reveals the possibility of surprise in his attempts to answer that question. If this is not me, then who is it, or what is it, and who am I? In the moment of estrangement Monette is unable to separate himself from the "not me," and his question opens up the possibility of encountering the "not me" in others as well as himself. Moreover, the movement "out of myself towards you in order to commune with that which ultimately I can never know" not only occurs across the spaces between bodies and between body and world, but also across time. Our experiences of ourselves are not contained or containable because, simply put, we experience them in time, and our knowledge of those experiences is always subject to time. We can begin to make this movement outside of ourselves towards difference when we acknowledge the possibility that *in time the not me might become me*, or, more simply, *in time the not me is me, is who I am*. Such a movement outside of oneself toward the other or the not me requires what Oliver calls vigilance, the next of my key terms from her work.

Vigilance

In a reading of Freud's *Beyond the Pleasure Principle* (1961 [1920]), Cathy Caruth notes that "[w]hat Freud encounters in the traumatic neurosis is not the reaction to any horrible event but rather the peculiar and perplexing experience of survival" (1996, 60). According to Caruth's reading of Freud, trauma is not simply the experience of a traumatic event itself, but the survival of that event; trauma is as much a "crisis of life" as a "crisis of death" (7). Survival is imbued with anguish because of one's traumatic encounter with death, and because one's own survival "is inseparably identified with victims who did not survive" (Langer 1991, 75).[9] In his writing, Monette demonstrates this peculiar and perplexing experience of survival in his vigilant witnessing to all those who did not survive. For Monette, the victims of AIDS in the early years of the epidemic include not only lovers and friends, but also countless others he will never know and whose voices—unlike his own—are now lost to us. Monette's own survival (for a time) makes him feel responsible for those countless lost voices, and his writing attempts to enact this response-ability, at the same time as it shows the immense difficulty of it.

In *Borrowed Time,* we learn that Monette in fact misses Roger's actual death;[10] he is sleeping "curled in Roger's bed" when the call comes from UCLA Medical Center. For Monette, sleeping does not avoid the fact of Roger's death so much as avoid the fact of his own survival; "waking," Monette writes, "teaches you pain" (1988b, 342). *Borrowed Time* ends with Monette postponing as long as he could "the desolate waking to life alone—this calamity that is all [his], that will not end till [he does]" (342). Freud discusses the intersections between sleeping, waking, and trauma in his analysis of the "dream of the burning child" in *The Interpretation of Dreams.* In this dream, Freud describes a father who falls asleep while his child, who has just died, lies in the next room watched over by a guardian who also, eventually, falls asleep. A candle next to the child topples over and sets the bed sheets and the child's arm on fire. The father is not awakened by the fire itself, but by a dream he has in which the child comes to him alive—and says "Father don't you see I'm burning?" (Freud 1965 [1900], 547–48). Freud suggests, according to Caruth, that the father not only dreams to keep the child alive, but also dreams so that consciousness as such does not wake up (Caruth 1996, 97). This is what Monette does as well in "Dreaming of You" from *Love Alone:* he dreams so that consciousness itself does not wake up. Lacan reinterprets Freud's dream by stressing the awakening as "*itself the site of a trauma,* the trauma of the necessity and impossibility of responding to another's death" (Caruth 1996, 100; emphasis in original). This necessity and impossibility of responding to another's death, and the repetition of this necessity and impossibility in dreams, is, according to Lacan, the missed encounter with the real—the *tuché*—which occurs "between dream and awakening, between the person who is still asleep and whose dream we will not know and the person who has dreamt merely in order not to wake up" (1977, 59). This necessity and impossibility of responding to another's death is defined "as an *ethical* relation to the real" (Caruth 1996, 92, 102; emphasis in original).

In "Dreaming of You," Monette elaborates further on the ways in which he attempts to hold at bay the nightmare of waking to his own survival, for in sleep and dreams Roger returns to him:

> . . . give me night
> give me more of it I wish to be an expert
> on darkness and all it conjures wish to sleep-
> walk with you no matter how queer a scene
> the crooked synapses of my brain cast us in
> a dream is never the one line long enough
> what's even worse we can't go walking after
> to watch from the canyon rim while the west
> burns midnight they are brief they are shadows
> they evaporate I wake I forget them
> but if they're all I have then let them come
> cascading . . . (1988b, 57)[11]

Sleep and dreams for Monette, however, do not "come cascading." In fact, in the essay "Sleeping Under a Tree" from *Last Watch of the Night* (1994), we learn that Monette suffers from acute insomnia, a condition which he presciently (or so it would seem) develops the night before Roger's diagnosis. Monette's insomnia, in its timing and manifestation, exemplifies what Oliver, following Emmanuel Levinas, defines as vigilance. As with the term "witnessing," Oliver notes two "radically different" meanings for vigilance: it is "both observing or keeping watch and responding to something beyond your own control" (2001, 134).

This double meaning of vigilance, as with the double meaning of witnessing, relies on an understanding of the Levinasian concepts of "beyond intentionality" and "wakefulness." On the one hand, the alertness and watchfulness of vigilance is "something that one intends to do," and on the other hand, it is "beyond intentionality" and "appears as a response to something or someone beyond one's self" (Oliver 2001, 134). In Monette's case, while

Roger is alive, his wakefulness has a purpose: he watches over Roger's much-needed slumber, preparing dosages at intervals throughout the night and feeding them to Roger "without really waking him up" (Monette 1994, 243). In the week of his death, Roger tells his doctor, "I'm sleeping for everyone now." The doctor understands his remark as a sign of "serious brain involvement," but Monette thinks the remark is "piercingly wise and tender," and believes that Roger is "sleeping *le sommeil du juste*—the sleep of the just—for all the rest of us, pursued day and night by our compromises with nightmares" (252). After Roger's death, Monette understands his role as the obverse of Roger's sleep for the just: "*I'm having insomnia for everyone now*" (252, emphasis in original). Monette's nocturnal vigils, in Oliver's terms, are "not the vigilance of a self-possessed watchfulness but the vigilance of a self opened onto otherness itself" (2001, 134). Monette's "self opened onto otherness itself" is the price he must pay for his own survival; it is, as he describes it, an exile into a "parallel universe, lunar and featureless" where all he desires is sleep.[12]

The call around 6 a.m. that awakens Monette after Roger's death is repeated again four years later when a call comes, this time just after 4 a.m., to say that Stephen, Monette's second lover to die of AIDS, is gone as well. "I think I've never stopped hearing that twice-tolled ring in the night," Monette admits, in the essay "Sleeping Under a Tree." He describes waking almost every night around 4 a.m. (having drifted off only an hour or so before) "in a panic, still waiting for that call. Sometimes the ghost of an echo, as if I've already missed it" (1994, 244). The call announces death, but he always misses it, hearing only "the ghost of an echo," which announces not death, but survival, and what Oliver describes as the "demands of otherness" (2001, 134). The vision of hope that sustains Monette and keeps him awake and writing until his own death comes in a dream he has while napping with Roger at the mouth of a secluded cave in Hawaii. Monette dreams of Kollau the Leper, who led a resistance movement against the American troops that sought to transport a group of lepers to a colony at Molokai. The American gunboats were unable to break the lepers' resistance, and eventually the lepers were allowed to stay put and create, what Monette calls, an "outpost of Eden and a tribe at peace" (1994, 260). The "memory of the dream encounter" becomes a touchstone for Monette, and he understands his vigilance, his determination to keep watch physically through his insomnia and figuratively in his writing, as the means by which he might create another outpost of Eden and bring peace to another tribe of others (261). Monette is the night watchman of his tribe of people with AIDS, and, as he tells his readers in "Sleeping Under a Tree," he will sleep only when he's dead.

Monette reveals over and over—and such repetition is inherent to the trauma as an essentially missed encounter with the real—that the trauma of Roger's death is also always the trauma of his own survival, and the ethical imperative that is inherent to that traumatic awakening. In the poem "Readiness," Monette considers suicide—"a cocked .32 will do in a pinch"—but admits, "I'm not half ready to leave us here without us all told" (14). The odd locution of this sentence points again to a death that is missed, yet still somehow must be recorded. In *Facing It*, Ross Chambers presents AIDS writing as an alternative—and the more difficult alternative, he believes—to suicide, which he describes as an "easier, and so tempting but ultimately unacceptable solution" (1998, 24). One must face death first as the appealing possibility of suicide, according to Chambers, before one can face death again—"in the form of living with, and dying of, AIDS" (24).

In the poem "The Very Same" as well as in *Borrowed Time*, Monette describes a moment just before Roger's death in which Roger, mostly blind, "sees" Monette come into his hospital room, and says, "*But we're the same person when did that happen*" (1988b, 20; emphasis in original). Roger's blindness is a form of seeing that entails more than vision. By "seeing" Paul as the same person as himself when he is blind and dying, Roger calls into question Monette's subjectivity. In doing so, he forces Monette to see himself from the vantage point of the other, expose himself to the other, and render account. "I had a self myself once but he died," Monette declares in the poem "Manifesto" (1988b, 41), and while it may be that that self died along with Roger, it may also be that that self died in Roger's blind vision of sameness that, paradoxically, requires Monette to remain open to the demands of otherness in his position between two deaths: Roger's and his own. This doubling—this death in life—that makes demands on Monette is apparent as well in the poem "Half Life,"[13] in which Monette grieves,

> . . . I get up and half of me doesn't
> work I drag me like a broken wing my good
> eye sees flesh and green the dead eye an X-
> ray gaping at skeletons . . . (1988b, 16)

Monette sees through both eyes—good and dead—and his writing presents both visions—flesh and green and skeletons. When Monette is told after Roger's death that it is "*time to turn/the page*," to move past Roger's death, he retorts, "BUT THIS IS MY PAGE IT CANNOT BE TURNED" (1988b, 20; emphasis in original). Monette's work can be distilled into a single page that cannot be turned and must be filled edge to edge with Roger and Paul's "growing interchangeability," until they are both "all told" (20).

Monette's survival gives him an ethical responsibility to bear witness to Roger's death and his own, and also to the death of a generation. He admits in *Borrowed Time*, "I can't help think of almost any moment of October [the month of Roger's death] without feeling helpless, like flinching in the glare of the final air burst. But how was I to know? Then I knew nothing about death, and now I know everything short of my own" (1988a, 323). This knowledge—knowledge not only *about* disaster but *as* disaster, as Blanchot would have it—is, paradoxically, both the impetus for and the impossibility of witnessing. Monette attempts to explain the paradox:

> It's like I died, and I *didn't* die. We are here, and we love each other, and now I have to find some work. Sentence by sentence, nothing by nothing, even if I can't sing. Then hum a few bars at least. Whistle a bit in the dark. We cannot all go down to defeat and darkness, we have to say we have been here. (1988a, 129)

"Sentence by sentence, nothing by nothing . . . we have to say we have been here." Both the sentences and the nothing are part of the knowledge and must also be a part of the practice of vigilant witnessing. Monette must find work that testifies to those who have been here and are now gone, as well as to love that is not perceived as such by the larger society. By humming a few bars and whistling in the darkness to say, "We have been here and we have loved each other," Monette counters what Oliver calls the "double alienation" of oppression, which "results not just from finding yourself in a world of ready-made meanings [which Oliver calls existential alienation] but from finding yourself there as one who has been denied the possibility of meaning making or making meaning your own without at the same time denying your own subjectivity" (2002, 56).

But when we say we have been *here* and that we have loved each other Monette despairs, will anyone *hear* and understand (consciously or unconsciously)? Will there be future witnessing? In his essay "Truth and Testimony," Laub notes that "the process of the testimony" is "essentially a ceaseless struggle" (Laub 1995, 61).[14] This is apparent in Monette's work, as he, in fact, writes about Roger's death right up until his own death: again, "I'm not half ready to leave us here without us all told." According to Laub,

> This imperative to tell and to be heard can become itself an all-consuming life task. Yet no amount of telling seems ever to do justice to this inner compulsion. There are never enough words or the right words, there is never

> enough time or the right time, and never enough listening or the right listening to articulate the story that cannot be fully captured in *thought*, *memory*, and *speech*. (1995, 63; emphasis in original)

The challenge for Monette as well as for us—as readers and as witnesses—is not only to tell but to listen; that is, to grasp, in Oliver's terms, "the unseen in vision and the unspoken in speech" (2001, 2). As I have attempted to show again and again, what is called for, what is demanded of the readers of illness narratives is a "new kind of listening, the witnessing, precisely, of *impossibility*" (Caruth 1995, 10; emphasis in original).

Working-through

What this new kind of listening might be, as embodied by Monette and as practiced in his writing, is theorized by Oliver in her delineation of—or perhaps, more appropriately, her working-through of—Freud's concept of working-through. In the poem "Three Rings" and again in *Borrowed Time* and yet again in the essay "3275" from *Last Watch of the Night*, Monette returns to two scenes. His generic searches are attempts to make language and meaning say what he needs to say, despite, or indeed because of, the double alienation that attaches to the experience of AIDS. What these scenes, and Monette's attempts to represent them, show is working-through as an ethical theory and practice. According to Oliver, "'Working-through' is a profoundly ethical operation insofar as it forces us not only to acknowledge our relations and obligations to others—that is, the ethical foundations of subjectivity—but also thereby to transform those relations into more ethical relations through which we love or at least respect others rather than subordinate or kill them" (2001, 68–69). Working-through, for Monette, will require that he bear witness—that he "love or at least respect others"—not just in and through language but also in and through his body, an experience which he then must attempt to render into language.

Monette describes such a process when he visits Roger's grave ten weeks after his death. "3275" is the number of Roger's grave, and it is also, as we realize at the end of the essay entitled "3275," the number of Monette's grave as well. Both Roger Horwitz and Paul Monette, then, are inscribed in Monette's writing, and in the inscriptions marking their grave.[15] In "3275" and in the poem "Three Rings," Monette tells of his visit to Roger's grave in which he buries in the grass a Zuni ring he has bought for Roger on a trip to New Mexico. After Monette buries the ring, he begins to moan, ventriloquizing Roger's own moaning in the hospital ten weeks before. The scene from "Three Rings":

> . . . suddenly I'm moaning out loud
> this very specific moan the echo of you
> when I walked in the last day a horn sound
> that knifes me still . . .
>
> . . .
>
> the moaning wouldn't go away so the day of
> the rings I mimicked you ventriloquizing
> your last sound desolate as a sea-bell
> trying to figure what the hurt was where
> had we disappeared to then I froze mid-moan
> saw it all in a blaze YOU WERE CALLING ME
> (1988b, 31–32; emphasis in original)

And from "3275":

> Now ten weeks later, a stillborn year before me, I finally understood that the bleating sound on that last day was Roger calling my name. Through the pounding in his head, the blindness and the paralysis, all his bodily functions out of control, he had somehow heard me come in. Had waited. And once I understood that, I went mad. My moaning rose to a siren pitch, and I clawed at the grass that covered him. Possessed with a fury to dig the six feet down and tear open the lid and clasp him to me, whatever was left. I don't even know what stopped me—exhaustion, I guess, the utter meaninglessness of anything anymore. (1994, 102)

I have quoted at length from these passages in order to give the reader a sense of the repetition of the scene both in Monette's life and in his writing. By mimicking Roger's moan, he embodies that moan; in his performance of the moan, it possesses him. He hears it not only with his ears and his brain, but with the tissues of his body; he becomes, in the words of Lawrence Langer, an "active hearer" (1991, 21).

His understanding is delayed, but even as he records the moment again and again in his work, he does not—cannot—record the process of that understanding in words. The moan is in excess of what can be understood in language,[16] but at the same time it must be brought to language in order that it may be heard:

> . . . I didn't know Death
> had reached your lips muscles gone words dispersed
> still you moaned my name so ancient so wild and
> lonely it took ten weeks to reach me now

> I hear each melancholy wail a roar like
> fallen lions holding on by your fingertips
> till I arrived for how many drowning hours
> to say Goodbye I love you all in my name (1988b, 32)

The moan is something greater than speech; it is the demand of otherness made across time and space. The moan becomes Monette's "weather and compass," giving his work both its essence and its direction. Like the Lacanian real, the moan is both that which lacks symbolization and that lack around which symbolization is formed. The moan *as moan* is impossible; for Monette, it must be given symbolization, it must become a name, his name, Paul. The moan must become Monette, and he it, and in this way it leaps across bodies and time. In the scene Monette describes, "emotions and affects migrate or radiate between human beings," as Oliver would say, through space and time, from Roger to Paul and beyond Paul to his readers. At the grave, Monette's moaning and digging are an attempt to get closer to Roger's death, and also to experience his (Monette's) own dying, his own burial, his own moment—an interminable moment, a moment outside of time—beyond the limit of language and knowledge.

In the working-through of this scene in various forms of writing, Monette reveals that, still, this moment is in excess of symbolization, it is beyond the recognition that the "hearing" of his name implies. To say that he finally and fully understands Roger's moan—that, in the grave scene, he grasps it—feels like consolation for a moment that seems to resist just such consolation. Like Sandra Butler's need to contain Barbara Rosenblum's mother's wailing at her daughter's funeral (not to silence it, but to contain it in her rendering of it) as a part of a "proper farewell" and a complete ritual (1991, 172), Monette's hearing of Roger's moan as his name is also a need to contain the force of that which is beyond language. Monette needs to represent Roger's moan in order to recuperate something he has missed and will always miss. He projects onto Roger's moan his own needs, but he also shows in that moment that he is saying more than he knows, doing more than he intended. Roger's death is a death that only Roger can experience. Monette is always going to be too late; he will always miss the call of otherness. He could not be a part of Roger's death because death is about both the absolute aloneness of the person who dies and the absolute aloneness of the person who survives. There is, therefore, in this moment of the moan a need that is never met. The performance of the moan says more than Monette's recognition of his name implies. Monette's interpretation of Roger's moan as his name is a need to contain the force of that which is beyond language.

By making such an assertion, I do not mean to imply that Monette is somehow wrong or mistaken in hearing his name in the moan, just as Butler is not mistaken in describing Barbara Rosenblum's mother's wailing as a proper farewell to her daughter. What I do want to suggest, however, is that in the witnessing of something beyond recognition, Monette transmits not only the fact of the moan in narrative and the translation of the moan into his name, but its affective force as well. We, the readers, must attempt to hear both the address and the affective force of the address; for through our own response-ability to both the address and the affective force of the address, we keep open the possibility of future witnessing. And keeping open the possibility of future witnessing is, finally, about love, another key term in Oliver's work that is a crucial aspect of a theory—and a practice—of subjectivity beyond recognition. Oliver writes, "To love is to bear witness to the process of witnessing that gives us the power to be, together. And being together is the chaotic adventure of subjectivity" (2001, 224). Monette's testimony to the experience of AIDS reveals this chaotic adventure of subjectivity in his performance of being together—with Roger, with countless others who have lived with and died of AIDS, and with his readers (even beyond Monette's own death of AIDS)—across space and time.

Past and Future Repetitions

In *Moses and Monotheism*, Freud says that latency is a feature of both infectious disease and trauma; for him, there is a gap—a latency—between the traumatic event and the appearance of symptoms (1967 [1939], 84). To take Freud's analogy full circle, HIV/AIDS is an infectious disease that also has the temporal structure of trauma. Latency, as Freud makes clear, is a characteristic of all infectious diseases, but HIV/AIDS has added a peculiar twist to this phenomenon of latency. This is partly because latency in HIV/AIDS has been such an unknown quantity; there is no standard amount of time from the point that one becomes infected with the HIV virus to the point that one becomes ill. Again, according to Monette, "No one has solved the puzzle of its timing" (1988a, 1). And, thus, as Sontag notes in *AIDS and Its Metaphors*, "[w]ith the most up-to-date biomedical testing, it is possible to create a new class of lifetime pariahs, the future ill" (1988, 33–34). With the advent of new treatment options, the experience of AIDS is, in fact, changing.[17] Nonetheless, I want to focus here briefly on the traumatic structure of AIDS, which has had individual as well as communal effects. "Has anything ever been quite like this?" Monette wonders. "Bad enough to be stricken in the middle of life, but then to fear your best and dearest will suffer exactly the same. Cancer and the heart don't sicken a man two ways like that" (1988a, 83). And as we've seen, prior to becoming sick, the person who is HIV+ often experiences—as friend, lover, and/or caretaker—the illness and death that awaits him or her; the PWA dies before dying, in between two deaths. In the gay community in the United States in particular as well as in communities across Africa and Asia, the numbers of AIDS deaths are unthinkable outside of wartime, which is one reason why Monette and others use military metaphors to describe the experience of AIDS. "Added to the caretaking and loss of others," Douglas Crimp maintains, "is often the need to monitor and make treatment decisions about our own HIV illness, or face anxiety about our own health status" (1990, 241; see also Kayal 1993, 9). This, of course, relates back to what I described earlier as the "crisis of survival," or the "crisis of life," which Crimp and others call "survivor's guilt," and which is inextricably linked to the "crisis of death," both in the past and in the future.

In the poem "Current Status 1/22/87," Monette explores the ways in which the present is a repetition of the past (1/22/87 is three months to the date after Roger's death), and a rehearsal of the future. Monette's poem describes his current *physical* status as of 1/22/87, including his latest T-cell count and the endless drug regimen he follows in the hopes of maintaining that count. Monette's doctors declare him *physically* fine, healthy, and asymptomatic; but *metaphysically* he is a wreck, a symptom only of Roger's death, in many ways already dead himself:

> my groping docs might just as well use leeches
> for all they can touch my invisible disease
> cracks on the heart don't flip on an EKG
> thus no treatment sorry we don't cure life (1988b, 35)

The crisis of Roger's death is the crisis of Monette's survival of Roger's death, and the crisis of a future repetition: "my condition is just a prefix," according to Monette (35).[18] His "groping docs" don't have any treatment for his current condition—the condition of his present survival and future death—because they can't actually see and record it with their instruments. For Monette, then, the set of printouts that declare him "clinically healthy," in fact, "sound like a qualification," and he wonders, "is this how being a hero starts or just dying?" (36). Although his work often resorts to images of a heroic time when men loved men and passed on knowledge in this way, in "Current Status 1/22/87" he senses what little compensation even a heroic imaginary is when one faces the real of "just dying." Lawrence Langer's rendering of the Holocaust survivor as a "diminished self" captures best, it seems to me, Monette's "current status." "Rejecting nihilism *and* heroism," Langer writes, "the diminished self lapses into a bifocal vision, as its past invades its present and casts a long, pervasive shadow over its future, obscuring traditional vocabulary and summoning us to invent a still more complex version of memory and self" (1991, 172). This bifocal vision emerges out of the experience of being between two deaths, and, in order to render such an experience in some form, one must invent a still-more-complex version of memory and self. This is Monette's "all-consuming life task," and, as we shall see, the invention of a more complex version of memory and self will become Bayley's task as well in the face of Iris Murdoch's Alzheimer's.

Bayley's Tragic and Comic Witnessing to Alzheimer's

Just as Monette's works are crucial in rendering the "experience" of AIDS in the United States in the 1980s, so too are Bayley's works in rendering the "experience" of Alzheimer's in the 1990s. Much about the experience of Alzheimer's had been, up until the publication of Bayley's books about Murdoch, hidden from history, and he makes visible a particular history that is at once personal, ethical, and medical. What is important about the rendering of experience, according to Joan Scott, is not simply to make visible that experience, but rather to show how experience is itself constructed, and to reveal the ways that experience is not a reliable or self-evident source of knowledge. Scott is concerned with how a particular social position—for example, homosexual, gay, queer—comes into being in a particular place and time. Memoir does not simply make this process visible, it is a part of the process of the emergence of a particular subject position itself. In the case of Alzheimer's, as with all disease entities, what it means to be a person with Alzheimer's—as opposed to, say, a senile person—is not self evident. The subject position "person with Alzheimer's"[19] has a history, and Murdoch's own Alzheimer's and Bayley's books about it (I am separating these two things here, but they are in fact very much interconnected) are not only part of a history of Alzheimer's, but also part of a history of the category "person with Alzheimer's." David Shenk's book *The Forgetting* makes this point too, at least implicitly. *The Forgetting*, as its subtitle suggests, is a "portrait of an epidemic," and Shenk historicizes the emergence of Alzheimer's as a diagnostic category in the 1970s. He notes that, although Alzheimer's was first observed and named at the very beginning of the twentieth century, it was not until the 1970s that "[a] critical mass of doctors began to *prefer* to see senility as a disease. The medical establishment was now ready to take on the moral challenge of Alzheimer's, to make a commitment to intervention" (2001, 133).

On February 8, 1999, Iris Murdoch died of Alzheimer's disease. She will be remembered as a great twentieth-century thinker and writer, as both philosopher and novelist. And yet, she will also be remembered for forgetting. As Murdoch sinks into the beyond of Alzheimer's disease, her husband, literary critic and novelist John Bayley, attempts to care for her and at the same time to write of their relationship as it was *then*, before Alzheimer's, and as it is *now*, after the appearance of Alzheimer's. His first book about their life together and Murdoch's life beyond the death of the self that occurs in Alzheimer's is divided into two parts: Part 1, "Then," and Part 2, "Now." The difference between past and present is formally marked as Bayley moves from numbered chapters in the first section to dated journal entries in the second section. Like witnessing to the experience and event of AIDS for Monette, witnessing to the experience and event of Alzheimer's for Bayley has a particular temporal structure. Alzheimer's transforms the form of Bayley's writing through its habit of imploding time into a present without past or future.

Although Alzheimer's has divided her—and his—life in such a radical way, the real of their present situation ruptures his attempts to describe their past together as serene and quietly loving. In portraying their marriage before Alzheimer's, Bayley emphasizes their capacity to feel close even—or especially—when they are apart. Once Alzheimer's descends on Murdoch, however, they are no longer able to find comfort and reassurance in the "solitariness each saw and was aware of in the other" (1998, 83). Bayley finds it difficult both to communicate with Murdoch and to be free of her, in spatial terms at least: "I sit at the table, and make desperate efforts to keep it as my own preserve, as it has always been. Iris seems to understand this, and when prompted goes obediently into the sitting-room where the TV is switched on. In less than a minute she is back again" (93). Now, both Murdoch's existential solitariness and, paradoxically, her inability to leave Bayley alone, makes *now* so radically different from *then*. "It is a way of life that is unfamiliar," Bayley writes. "The closeness of apartness has necessarily become the closeness of closeness" (92). Bayley's rendering of this transformation of the closeness of apartness becoming the closeness of closeness might be a definition of Freud's uncanny: what is familiar becomes so familiar it is strange.[20] As we read in *Iris*, the real Iris disrupts Bayley's attempts to write about her; his words become muddled because he must be attentive to the real Iris, not the Iris of the past and in his writing (94).

For Bayley, as for Monette, one book, it seems, cannot capture or contain his mourning over Murdoch's diminishment and then death: again, in Laub's words, "the imperative to tell and be heard become[s] itself an all-consuming life task" (1992, 63). Murdoch is still alive at the end of the first volume, which seems to necessitate (along with its enormous and surprising popularity) the second volume, which contains a description of end-stage Alzheimer's and Murdoch's final days. The second volume also contains much about Bayley's childhood and life before Murdoch. He explains that Murdoch's loss of memory—and her concomitant loss of self—has motivated him, paradoxically, to remember and to delve into those memories that have been formative to the cultivation of his own self. Although this explanation makes sense to me, his second volume, nonetheless, reads as though he feels he can now take advantage of the popularity of the first volume, and his increased fame as a result of it, as an excuse to write about himself without Murdoch. Or, perhaps, the trilogy itself is an attempt to return, through narrative at least, to the closeness of apartness that had been lost with Murdoch's Alzheimer's.

Bayley's own self re-emerges, or is re-fashioned, in the memoirs. Murdoch, and the memory of her along with her lost memories, motivates Bayley to write these memoirs, but ultimately they are about Bayley as much as—if not more than—they are about Murdoch. What is cultivated, especially in the second and third volumes, is Bayley's self, or at least a fantasy self. Even after the first volume, there were some who protested Bayley's portrayal of Murdoch with Alzheimer's. The release of each new volume seems to have increased the numbers who find Bayley's work self-indulgent at best and reprehensible at worst. Although I do not have the space here to take up in detail the question of appropriation, I do want to at least suggest again that this question is still very much operative with regards to both Monette's and Bayley's work. Because the second and third volumes of the trilogy move away from attempting to represent Murdoch and her Alzheimer's, I concentrate in particular on Bayley's first book in my discussion of his work in this chapter.

But, briefly, the backlash against Bayley: In *Widower's House* (2001), Bayley describes his experiences as England's most famous widower, who falls prey to numerous charitable women who seek to comfort him in his grief. Two women in particular (one in her early thirties, the other older, but still much younger than Bayley), whom Bayley calls, allegedly for the sake of discretion, Margot and Mella, invade Bayley's house (in order to cook and clean for him) and his bed. Calm is only restored to the widower's house when he marries Audi Villers, who, with her husband, Borys, had been Murdoch and Bayley's closest friends. Bayley describes his marriage with Villers as an almost foregone conclusion of the foursome becoming a twosome after the deaths of Iris and Borys.

There is little doubt that Bayley has gotten a lot of mileage—or, pages and column inches, as the case may be—out of Murdoch's Alzheimer's.[21] It is unclear whether his work on his life with Murdoch was originally intended as a trilogy, or whether the popularity of the first and/or his continued need to write as a "good therapy for grief" and as a "form of prayer," as he described his writing in two separate essays in the *Daily Telegraph*, has necessitated this form (2000b, 21; 2000c, 22). Working in January of 2000 on *Widower's House*, Bayley admitted, "I can't stop writing about Iris, although this book will have to be about living without Iris" (2000b, 21). But, to what extent *Widower's House* is actually about Iris is debatable. I mention these latest developments in the saga surrounding Bayley and Murdoch not to cast doubts on the sincerity of his "inner compulsion" (Laub 1992, 63) to write about her, but rather to note that Bayley seems to have been transformed as a particular kind of writer as a result of Murdoch's Alzheimer's and his writing about it. After the publication of volumes one and two, Bayley was seen by many as an expert not only on Alzheimer's but also on love and marriage. After *Widower's House*, Bayley fashioned himself, according to John Sutherland in the *Guardian*, as something of a septagenarian "sexual athlete" (2001, 4). And the question remains, where in Bayley's last memoir and in the publicity surrounding its rather lose articulation of the "truth," is the Iris he supposedly can't stop writing about? Like Scott, I want to problematize the practice of memoir-writing and its relationship to experience in general. Bayley's trilogy, as both a cultural and a literary product, exemplifies the impossibility of bringing "experience"—of the self and others—to language outside of available literary and cultural forms of representation. It also problematizes again the difficulty of speaking for another, especially when that person has lost the capacity to speak for herself because of a debilitating disease like Alzheimer's. But Bayley's appropriation of Murdoch's story in the form of memoir is especially interesting, and perhaps especially troubling, in light of Murdoch's own attempts to delineate an ethics that sought to diminish the narcissistic self, and it is to Murdoch's ethics that I now want to turn.

Tragic Witnessing

Before I look further at Bayley's writings about Murdoch and her Alzheimer's, I want first to discuss briefly Murdoch's moral philosophy, and, in particular, her understanding of the ethical act of "unselfing" (1992, 54). Or, more precisely, I want to read Bayley's memoirs with and against Murdoch's philosophy in order to consider the ways in which, together, they delineate a conception of the comic and tragic, which might help us to understand witnessing as an ethical practice. As I mentioned at the outset of this chapter, many have seen Murdoch's demise as a tragedy, but is it a tragedy in Murdoch's understanding of the term? According to Murdoch's literary and philosophical treatments of the concept and genre of tragedy, tragedy "must break the ego" (104). Alzheimer's is nothing if not a breaking of the ego. But what, then, is Bayley's trilogy of Murdoch's experience of Alzheimer's: is it a description of that breaking and broken ego, or something else? Is it possible for Bayley's memoirs to put into language Murdoch's experience as she transgresses the limits of the self, and becomes a self unselfed, beyond itself? After all, Bayley, unlike Monette, will not—inevitably—die in the very same manner as Murdoch. In attempting to describe the experience of Alzheimer's, he can only describe it from the outside,[22] which, to some extent, is a phenomenon peculiar to Alzheimer's and other forms of dementia. Just as Matt Seaton came to the painful realization that *his* experience of his wife's death from breast cancer that had metastasized to her brain was not at all the same as—or, even, despite their proximity to each other, proximate to—*her* experience, Bayley himself does not transgress the limits of humanness—he is not between two deaths—but he does attempt to communicate with Murdoch beyond that limit, and to describe to his readers the form in which that communication takes. And because he understands himself fundamentally through his relationship to Murdoch, when she is experiencing the "inarticulate experience of the inside" of her Alzheimer's, as Oliver might say (2001, 90), he is in a sense *in it* too. Thus, it is Bayley's treatment of the transformation of language and its relationship to the transformation of self in Alzheimer's that I am interested in here.

In her essay "Comic and Tragic" in the rather eclectic collection of philosophical essays *Metaphysics as a Guide to Morals* (1992), Murdoch attempts to delineate the relationship between the comic and the tragic.[23] She insists that they are not precisely opposite, but "asymmetrical and different in kind" (92). This asymmetry, for Murdoch, does not have to do with their relationship to seriousness (both can be serious), but rather with their relationship to the distinction between art and everyday life. The comic, according to Murdoch, is everywhere, it is in the air we breathe; we are all capable of comedy, in fact, our very being is comic. The tragic—or, more specifically, tragedy—"belongs only to art, where it occupies a very small area" (92). There are, Murdoch believes, very few good tragedies, and bad tragedy is not tragedy

(92–93). Although her claim is perhaps an elitist one, more importantly, she wants us to consider the relationship between art and real life, between literary form and the real. Although Murdoch does not take up the memoir form in her work, she does imply that it might not be a form well-suited to tell of pathos. For Murdoch, "[r]eal life is not tragic"; rather, "[s]trictly speaking, tragedy belongs to literature," and, in particular, to great literature (93).

Is she, then, making an aesthetic judgment about what is suitable for telling and in what form it should be told, as Croce does in her polemical piece that I discussed in my introduction? I don't think so. Rather, her claims are about the real (and the Murdochian real is surprisingly close indeed to the Lacanian real) and the impossibility of telling it: "Real life is not tragic. In saying this one means that the extreme horrors of real life cannot be expressed in art" (93). This is because art, for Murdoch, always "offers some consolation, some sense, some form, whereas the most dreadful ills of human life allow of none" (93). I am concerned here with the way illness narratives attempt to make sense of and give form to some of the most dreadful ills of human life, and the belief that both writing and reading such stories of "real life" can offer consolation. And yet, I am also interested in the fact that such attempts to make sense, give form, and offer consolation often also fail. What might this failure signal in terms of an ethics? What happens when the structure of consolation breaks down and something else—beyond consolation—is glimpsed? It seems to me that consolation is a conservative gesture, not an inventive one; it implies a comfort in old, familiar forms, not a desire for new ones that cannot offer consolation.

I suspect Murdoch would view the proliferation of memoirs in contemporary culture with weariness at best, not necessarily because it is a popular, democratic rather than a high literary form, but because it is a form that often privileges the cultivation of a unified self. Although she recognizes the possibility of art to give us "glimpses of another scene" (36), she is skeptical of our habit of encountering art as "a spurious short-cut to 'instant wisdom'" (19).[24] Art as instant wisdom does not challenge the unity of the self; it simply and immediately gives the self what it wants, or what it thinks it wants. It is a quick fix, not a treatment in which the self is called into question by the demands of otherness rather than taken for granted. According to Murdoch,

> The tragic art form is rare because it is difficult to keep attention focused on the truth without the author slipping into an easier sentimental, abstract, melodramatic (and so on) mode. . . . Death threatens the ego's dream of eternal life and happiness and power. Tragedy, like religion, must break the ego, destroying the illusory whole of the unified self. (104)

Most memoirs, even memoirs of illness, do not focus on the breaking of the ego, but rather show the ways in which the ego maintains itself in the face of its own dissolution. For Murdoch, great tragedy does not give us redemption, but death unredeemed, as in *Antigone*,[25] or as in Murdoch's exemplary case, *King Lear*. Tragedy must resist the comforts of consolation and redemption; it must be, Murdoch writes, "in a positive, even thoroughly uncomfortable sense a broken whole" (104). She acknowledges, nonetheless, that as art, tragedy "must have borders, [and] it must be some kind of magic." But, she continues, it "must also inhibit magic in its familiar and consoling uses" (105). What tragedy must hold on to, according to Murdoch, is the "pain of contingency, which is a shadow of death" (111). This, it seems to me, is ideally what illness narratives must hold onto as well, though most inevitably fail to do so, because the pain of contingency and the shadow of death are kept (of necessity?) at a distance. Simply put, in most illness narratives, it is easier to maintain the limits of language and death than to explore beyond them.

In her discussion of *King Lear* as an exemplary tragedy, Murdoch reads Lear's death at the end of the play as a kind of relief, providing a wished-for peace for the play's spectators. She insists that a far more terrible—and more real—ending would have him not dying but living on in utter anguish with his crying continuing without relief; such would be an "image of death which would be more awful than the merciful end conventionally required" (120–21). Murdoch wonders how we would bear this awful image of death, this unending demand of otherness. There is something in this image that recalls Roger Horwitz's moaning, and that also might inhibit the kind of consoling magic that insists on making his strange and unbearable moaning familiar, and, therefore, more bearable, by allowing Monette and us to finally comprehend it. This image isn't nearly so consoling—isn't consoling at all—if Monette must continue to ventriloquize it without it ever reaching him in a language he can understand. That is, what if it reaches him and continues reaching him, but he *never knows its meaning?* That is the pain of contingency of which Murdoch speaks. While Monette's writing does reveal this pain of contingency, at the same time, he must cover over it with a consoling certainty that he has understood what he hears. Murdoch's image of Lear crying on and on into the future is not only useful juxtaposed with Monette's image of Roger's moaning, it is also useful juxtaposed with Bayley's image of Murdoch's Alzheimer's. Is Bayley's portrait of Murdoch's Alzheimer's a tragedy in Murdoch's sense of the word? Does it "inhibit magic in its familiar and consoling uses"? Does it manage to give us a glimpse of the pain of contingency and the shadow of death that is Alzheimer's? Or, does it seek only to console him and us?

Bayley's first memoir of Murdoch ends on Christmas Day 1997, and with the consolation of the "Christmas ritual" for him and also seemingly for Iris. As he comes to the end of his memoir of her, Bayley notes that, "Iris is listening, which she rarely does, and smiling too. There have been no anxious pleas this morning, no tears, none of those broken sentences, whose only meaning is the dread in her voice and the demand for reassurance" (1998, 189). Iris listening and smiling on Christmas Day comforts Bayley, despite the anxious pleas, tears, broken sentences, and dread that he faced earlier. Iris listening and smiling does not make demands on Bayley that he cannot meet nor even fully comprehend. Bayley admits that in this moment he is under the "illusion . . . that life is just the same, has never changed. I cannot now imagine Iris any different. Her loss of memory becomes, in a sense, my own" (189). He recalls a comment from Wittgenstein, quoted to him by Iris, "that death is not a human experience. We are born to live only from day to day" (189). This quote, relayed from Wittgenstein to Bayley through Murdoch, comforts Bayley, but such comfort seems far removed from Lear's unending cry imagined by Murdoch. Yet Wittgenstein's notion that death is not a human experience is performed in the text of Bayley's second memoir. The book moves inexorably toward Murdoch's death, but doesn't, or can't, present it. Although the reader has been expecting her death, we somewhat abruptly and rather obliquely learn of it only in the epilogue. What stands in for Murdoch's death at the end of *Iris and Her Friends* is Bayley's loss, first when he puts her in a nursing home, and then after her death. Bayley tells us that death is the "best of friends" for Murdoch because it brings her, finally, reassurance. But is death a friend who brings reassurance to the person with Alzheimer's or to her caregiver and keeper of her memory?

Bayley's task in his trilogy is to describe what happens—to Murdoch, of course, but also to him and to their relationship—with the onset of Alzheimer's, but, in describing the losing that occurs with Alzheimer's, he must also describe that which has been lost. Murdoch is an internationally known public figure, but her fame does not mean that the public knows her in any intimate way. What Bayley hopes to offer his readers, then, is an intimate portrait of Murdoch: Murdoch as companion in their day-to-day life together, rather than, or as well as, Murdoch as novelist and philosopher. But, at the same time as he provides an image of Murdoch as more than philosopher and novelist, he must also provide an image of her as much less than this. Murdoch's Alzheimer's is first revealed to Bayley when they are both participating in a conference in Israel. Murdoch has agreed to answer questions from the audience about her work, but it soon becomes clear to both Bayley and the audience, though seemingly not to Murdoch herself, that she is unable to perform this task.

This is Bayley's first glimpse—his "discovery of knowledge," as Laub might say—of the diminishment of her public language, and it prompts him to recall a recent conversation with her about the novel she is working on at the time, which would become Murdoch's last published novel, *Jackson's Dilemma* (1995). Bayley explains that Murdoch often expressed frustration over a novel in progress, but would very quickly get the better of her frustration and push on through with her writing. "[T]his time," Bayley writes, "it was quite different." With "worried detachment," Murdoch tells him, "It's this man Jackson . . . I can't make out who he is, or what he's doing." When Bayley attempts to ease her worry with a joke—"Perhaps he'll turn out to be a woman," he says—she responds in a very serious manner, "I don't think he's been born yet" (148). It is as though, through a character in one of her novels and through the practice of writing itself, Murdoch herself gets a sidelong glimpse of the real that is happening to her.[26] Bayley misses the meaning of this conversation at the time, misses the fact that, perhaps, he is communicating with Murdoch from across the limit of the death of a self, or a self—like "that man Jackson"—that hasn't yet been born. In *Iris and Her Friends*, Bayley returns to this conversation and to the subject—"that man Jackson"—of Murdoch's last novel. Bayley recalls again that Murdoch was puzzled by her character, and couldn't figure out who he was. Bayley retells this story because now it seems obvious to him who Jackson was: "It seems obvious today—perhaps too obvious?—that in one sense at least, he was what was coming. The dark foreshadow of her present disease" (2000a, 63–64). This is the stuff of Murdochian tragedy, but it is also, as we shall see, the stuff of comedy.

Monette's question—"How is this not me?"—is a pertinent question for Bayley as well: "How is this not Iris?" That is, how is Alzheimer's not Iris, or what of Iris remains behind or within the shell of Alzheimer's? Bayley tries to show us the ways in which Murdoch is transformed by Alzheimer's into someone else, but also the ways that she—her *self*, or some part of herself—is unchanged by it. "Alzheimer's sufferers are not always gentle: I know that," Bayley states simply. It is not that he denies a new knowledge of Murdoch that Alzheimer's reveals. "But," he also insists in the very next sentence, "Iris remains her old self in many ways" (1998, 34). When an acquaintance of Bayley's, whose husband also suffers from Alzheimer's, tells him, "cheerfully," that looking after someone with Alzheimer's is "like being chained to a corpse," he does not agree with this woman's comment—he is, in fact, "repelled" by it (42). He recognizes instead that "[o]ne needs very much to feel that the unique individuality of one's spouse has not been lost in the common symptoms of a clinical condition" (42). Bayley's portrait of Murdoch is meant to reveal the ways in which she is not a corpse (even if or because she is between two deaths), the ways in which she is both fundamentally transformed and yet still Iris.[27] And yet, in showing the ways in which she remains her old self, he must first also show what has been lost:

> The power of concentration has gone, along with the ability to form coherent sentences, and to remember where she is, or has been. She does not know she has written twenty-seven remarkable novels, as well as her books on philosophy; received honorary doctorates from major universities; become a Dame of the British Empire. (34)

Bayley describes how, when Murdoch is asked by an admirer of her work to sign one of her books, although she obliges and laboriously signs her name, she has no understanding of what this signature means, or its connection to who she is (or was): that is, she has no knowledge of her self as author. Murdoch the author is dead; nevertheless, Murdoch the person with Alzheimer's is "always anxious to oblige." According to Bayley, "[i]t takes her some time, but the letters are still formed with care, and resemble, in a surreal way, her old handwriting." All that remains, Bayley notes, is "the old gentleness" (34). Her new self resembles her old self, but only *in a surreal way*.

Comic Witnessing

As Bayley discovers, the best way to communicate with Murdoch's transformed self is also in a surreal way. Via the absurd and the nonsensical—the comic—he is able to maintain in the present an intimacy with Murdoch across the limit of Alzheimer's. At the same time, this surreal communication enables him to remember—to hold onto—a past self not wholly cut off from her present self, not wholly lost to Alzheimer's. It is the comic, then, that offers some consolation for the tragic, at least in the beginning: "A burst of laughter, snatches of doggerel, song, teasing nonsense rituals once lovingly exchanged, awake an abruptly happy response, and a sudden beaming smile" from Murdoch (43). Bayley goes on to compare such forms of communication to an encounter in the past "between explorers and savages, when some sort of clowning pantomime on the part of the former seems often to have evoked instant comprehension and amusement" (43). Revealingly, Bayley encounters Alzheimer's as if he were encountering a "savage," a primitive being that, in the case of Alzheimer's, does not dwell in another country at a distance that must be traveled to, but, rather, dwells close to the self as the "proximate other."[28] In seizing upon such a metaphor, Bayley seems to desire, like the explorer, to observe closely and yet also maintain his distance from the "savage." And yet, at the same time, that otherness is so close because, as he discovers, it is not outside of him but a part of him. He is able to maintain intimacy with Murdoch, then, not by having a rational conversation with her, but by taking up the form if not the content of Murdoch's animated exchanges, by allowing the irrational in himself to express itself to her. Bayley comes to understand that, "[o]nly a joke survives, the last thing that finds its way into consciousness when the brain is atrophied" (45).

As Murdoch's capacity for public language has been forgotten, what remains is a private language that they share, which, as Bayley notes, "cannot get us far," at least not in the public realm, but can nonetheless get beyond the limit that separates Bayley from Murdoch, life from death. "Knowingness," as Bayley calls it, will not do; only unknowingness and a certain *unconscious communication* will work and allow him to move outside himself toward Murdoch. I take the phrase "unconscious communication" from British psychoanalyst Christopher Bollas, who notes that this form of communication "is not necessarily about constructing lucid, effective, and memorable understandings of one another; rather, it is a way of life—for people in fairly constant proximity to one another. One person's direct effect on the other—unconscious to unconscious—cannot be witnessed by consciousness" (1995, 28). What becomes difficult, then, is to communicate this unconscious communication itself on to others in narrative form. Bayley attempts to be both a conscious and an unconscious witness to the "experience" of Alzheimer's, and the ethics that emerges from unknowingness rather than knowingness is an inventive ethics that is less concerned with recreating the past, and more concerned with inventing an unthought future.

In his memoirs, Bayley wants to mediate this unknowingness, to pass on this unconscious communication between Murdoch and him to beyond them, to those who will read these works. Bayley's works about Murdoch are, therefore, affective histories, in the sense that they seek to describe the private language he and Murdoch develop, as well as effective histories, in the sense that they also seek to communicate an experience of Alzheimer's that has not yet been told in a public language and beyond the private realm of the family. Thus, he supplements the science and medicine of Alzheimer's with a story of the effects of the illness on the person with Alzheimer's and on those around her. Like Monette, with his embodied comprehension of Roger's call to witness, Bayley very much wants to show that, despite the boundaries that separate him from Murdoch, he, nonetheless, can become an "active hearer," most often through humor and a refined sense of the absurd. And yet, what is clear as well from his writings on Murdoch, is that this form of communication isn't much compensation for "that endless omnipresent anxiety of Alzheimer's, which spreads to the one who looks after the sufferer" (1998, 35), as a call to otherness that cannot be apprehended. The incapacity of the caregiver to provide comfort to the person suffering from Alzheimer's means that often all that manages to be communicated from beyond the limit of Alzheimer's is a free-floating anxiety, a "perpetual query" that remains both over- and under-articulated. "An Alzheimer sufferer begins many sentences," Bayley explains, "usually with an anxious repetitive query, but they remain unfinished, the want unexpressed" (43). Often, Murdoch's perpetual query leaves Bayley's "own mind and memory faltering, as if required to perform a function too far outside their own beat and practice" (43).

Bayley must negotiate a method by which he might perform certain functions that seem, at first, to be too far outside the rhythms and practices of his own mind and memory. He negotiates this method in and through his writing, but, at the same time, this method inevitably fails because his writing cannot get at the thing itself—the meaning of and from within Murdoch's Alzheimer's. Bayley wonders, therefore, if "Iris speak[s], inside herself, of what is happening?" And, he admits, simply, "How can I know?" (172). As with HIV/AIDS, Alzheimer's is an epistemological crisis, but is it one for Murdoch, as it seems to those of us outside of her predicament? Is her perpetual query a sign of an epistemological crisis *for her*, or the sign of something else—something we cannot know? Murdoch's perpetual query leads to Bayley's own query: "How can I know?" This is the crucial question for him and his readers: how can he know anything about Murdoch's "experience" of Alzheimer's? And, moreover, what does this failure to know teach us about "experience" itself, and the attempt to bring this "experience"—any experience—to language? In the end, then, their mode of communication, he tells us, "seems like underwater sonar, each bouncing pulsations off the other, and listening for an echo" (44). To some extent, this is what all communication is: we can't hear—and know—the thing (the "experience," the real) itself in the moment, we can only listen for its echo, its effects in time. Rendering "experience," then, is not only about rendering the past, but also about rendering possible futures in and through language. By exposing the "limits to representation and representability," what emerges is not the impossibility of ethical action, but the necessity of creating new forms—both comic and tragic—of public and private language.[29]

Conclusion

Toward an Ethics of Failure

I end *Treatments* by beginning to formulate what I call an ethics of failure: that is, an ethics that emerges out of, or along with, an experience of failure, be it of the body, of (conventional and alternative) medicine, or of language. I explore the ethical implications of failure in a final case study, which treats two illness narratives: one, Atul Gawande's *Complictions* (2002), from the doctor's side of the doctor-patient binary, and the other, Gillian Rose's *Love's Work* (1995), from the patient's side. In order to delineate the ethics of failure that Rose's and Gawande's works suggest, I turn to Elaine Scarry's phenomenological discussion of the experience of pain, which she takes to be an experience that unmakes the world of the person in pain,[1] and Lyotard's exposition in *The Postmodern Condition* (1984) and *The Differend* (1988) of modes of presenting the unpresentable. Both Scarry and Lyotard pay attention to forms of affect that exceed the capacity of representation, and both seek methods to bring into being new "idioms which do not yet exist" as an ethical imperative in the face of suffering (Lyotard 1988, 13). These new idioms can be either aesthetic or political, or sometimes both.

It may seem counterintuitive to end a book about illness narratives with failure and an ethics that emerges out of failure. I am not simply trying to be provocative, although I do want *Treatments* to provoke a response within the spaces and relationships in which the experiences and events of illness come into being. My motivation is not unlike that of critic Arlene Croce a decade ago when she discussed "the undiscussable" in her review of Bill T.

Jones's dance performance, which included the words and images of real people talking about their experiences of illness. Jones brought art, politics, and illness together, and Croce wasn't moved by the conjuncture. My concern with the problem of appropriation throughout *Treatments* reveals an impatience—also not unlike Croce's—with some of the claims made by the writers of illness narratives concerning what art about illness can and should do. Nonetheless, I still want to forcefully argue against what I see as Croce's nostalgic attempt to keep art, politics, and personal experiences of illness in separate domains whose boundaries are maintained by the supposed experts of those domains. Croce's piece plays a central role in my cultural study of illness narratives because it signals the breakdown of the boundaries between various domains, and even the possibility of a reconfiguration of the category "expert" where illness is concerned. Her piece also suggests the anxiety such a breakdown and reconfiguration causes. Croce's own anxiety is with the position of the critic as expert on art, not politics or illness, and the "undiscussable" for her is that certain works that pass themselves off as art are not art, but present the experience of illness unmediated. In a sense, Croce's claim that art should be disinterested is not unlike the claim that science should be disinterested. Croce's view of the critic is as a kind of aesthetic clinician applying the anatomo-clinical method to art, and her diagnosis is that art in the age of AIDS is sick, especially when it attempts to present to its audience something about the experience and event of AIDS as lived now. For Croce, art must transcend the specificity of the moment of its emergence, not grapple with this moment in all its immediacy, intimacy, and messiness.

I am concerned with a different undiscussable: that doctors and patients might and often do get things wrong, and that they might not have a language or an ethics with which to communicate or transform this feeling of getting it wrong. Central to medicine, and to most illness narratives, is the doctor-patient relationship, and central to that relationship, and again to most illness narratives that often describe an experience of that relationship, is what Lyotard would call a *différend*, which he defines as an "unstable state and instant of language wherein something which must be able to be put into phrases cannot yet be" (1988, 13). Throughout his work, Lyotard identifies an infinite number of "phrase regimens," which are various "mode[s] of presenting a universe" that are "incommensurable" and not translatable from one mode to the next (128). Despite this fundamental incommensurability, Lyotard understands that the failure to link phrase regimens might lead not to paralysis or further suffering (though it certainly might lead there), but to a "lot of searching . . . to find new rules for forming and linking phrases that are able to express the differend disclosed by the feeling" (13). The hyphen that separates the two subject positions in the phrase "doctor-patient relationship" signals a *différend* in the doctor-patient relationship. What happens when both sides begin searching for new rules for forming and linking phrases between these subject positions? I have explored this searching through various practices of politics and witnessing, and I want to conclude with two final cases that suggest failure might be the beginning of such searching, not its conclusion.

Gawande's Imperfect Science

In *Complications*, Atul Gawande, who at the time of writing this series of linked autobiographical essays was a surgical resident in Boston and staff writer on medicine and science for the *New Yorker*, presents medicine as a fundamentally "imperfect science," because it is "an enterprise of constantly changing knowledge, uncertain information, fallible individuals, and at the same time lives on the line" (2002, 7). Although he assures his reader that in medical practice "[t]here is science in what we do," he also acknowledges that along with science there is "habit, intuition, and sometimes plain old guessing" (7). In order to explore how medicine is practiced day to day, "the moments in which medicine actually happens" (7), as Gawande puts it, he recognizes that it is not enough to merely investigate what medicine knows, but that he must also consider what it doesn't know, and how it "grapple[s] with that ignorance more widely" (8). Gawande's investigation into what medicine doesn't know is framed by three key terms that, along with imperfection, are not usually associated with twenty-first-century Western medicine: fallibility, mystery, and uncertainty. What happens when we acknowledge medicine's fallibility, mystery, and uncertainty rather than try to maintain the myth that medicine is all-knowing and all-powerful?

Although, as we saw in chapter 4, Verghese in particular recognized the impotence of medicine when confronted with a new infectious disease at a moment when Western science thought it had mastered infectious diseases,[2] Gawande makes a larger claim: He contends that fallibility, mystery, and uncertainty are characteristics of modern, Western medicine in general, not just in the particular case of emergent and deadly infectious diseases.[3] To suggest this possibility necessarily *complicates* our understanding of medicine and its practices. According to Gawande, in medicine "[t]he gap between what we know and what we aim for persists. And this gap complicates everything we do" (7). Gawande's "empirical philosophy"[4] of this gap—his close analysis of the way medicine is practiced, not what it claims to know—is an ethical project that emerges from what medicine doesn't yet know; it is an ethics of failure.

To insist on the absolute authority of medicine and to deny that doctors make mistakes or that they sometimes use intuition not science doesn't close the gap between what we know and what we aim for. Knowingness, as Bayley would call it, closes off rather than opens up routes to more effective treatments. Gawande presents an unusual view of medicine and its practitioners. He explores the day-to-day realities of how new and established doctors are trained, what happens when doctors make errors, what tools they have or don't have to understand and treat conditions, like pain and nausea, that are exceedingly difficult to treat, and how doctors and patients arrive at treatment decisions together. Gawande's view of medicine is unusual because, as Susan Wendell notes in her important feminist treatment of disability, *The Rejected Body*.

> Collectively, doctors and medical researchers exhibit very little modesty about their knowledge, rarely admitting to patients or the public the vast remaining gaps in scientific medicine's understanding of the human body or their inability to repair or heal most physical conditions that cause suffering, limitation, and death. Scientific medicine participates in and fosters the myth of control by focussing overwhelmingly on cures and lifesaving medical interventions, and by tending to neglect chronic illnesses, rehabilitation, pain management, and the quality of patients' experiences, including their experiences of dying. (1996, 94)

Gawande is indeed modest about what Wendell calls medicine's "cognitive authority," and *Complications* offers a new idiom, not of "medicine's twenty-first century, tall-in-the-saddle confidence" (2002, 193), but of "how messy, uncertain, and also surprising medicine turns out to be" (4). Rather than emphasizing Western medicine's amazing accomplishments over the past century, though he (and Wendell) don't deny that there have been many, especially in its ability to treat traumatic injury, Gawande discusses the undiscussable in medicine, what I call, following Lyotard, medicine's *différends:* including its frequent errors and continuing ineffective response to suffering caused by pain and nausea.

Errors

Complications begins with fallibility, medical education, and the obvious though rarely expressed point that medicine is a practice that requires certain skills, which are constantly changing as new technologies and protocols are developed, that can only be acquired through practice. It begins in particular with an example of Gawande's own fallibility and medical education: as a new surgical resident, he has a difficult time learning how to insert a feeding line into a patient's chest. The first time he tries, he fails, and eventually the senior resident has to take over to complete the procedure, but not before he inflicts some pain on the patient. Gawande explains that such incidents in which a doctor gets a procedure, treatment, or diagnosis wrong are frequent in medicine, partly because new doctors—and even experienced ones—have to learn new skills.[5] At the same time, this "learning is hidden, behind drapes and anesthesia and the elisions of language" (24). And if the learning itself is hidden, the errors are even more so. The disappeared learning, and the errors that come with it, appear only obliquely, when, for example, "an attending physician brings a sick family member in for surgery, [and] people at the hospital think hard about how much to let trainees participate." Gawande continues:

> Even when the attending insists that they participate as usual, a resident scrubbing in knows that it will be far from a teaching case. And if a central line must be put in, a first-timer is certainly not going to do it. Conversely, the ward services and clinics where residents have the most responsibility are populated by the poor, the uninsured, the drunk, and the demented. Residents have few opportunities nowadays to operate independently, without the attending docs scrubbed in, but when we do—as we must before graduating and going out to operate on our own—it is generally on these, the humblest of patients. (24)

Although I find Gawande's reference to the patients most likely to be practiced on as "the humblest patients" an elision of language in its own right (isn't "powerless" more to the point?), his delineation of who is and isn't likely to be practiced on reminds us of the importance of what I call a person's health capital in relation to getting good care. There is no doubt that certain personal relationships can provide an individual access to good care that another individual without those same relationships will not have access to.[6]

Gawande wonders what would happen if we recognized that mistakes are not aberrant, but fairly common in medicine (47), and then closely analyzed how and why, when and where they occur. Again, he gives us an example from his own experience as a surgical resident. This time he recounts a case in which he is unable to perform an emergency tracheotomy; his failure to do the tracheotomy almost has very serious consequences for the patient, but fortunately, thanks to a rather miraculous save from an anesthesiologist who manages to intubate the patient just in time, everything ends well. Gawande presents this incident with candor: he admits he was "ill-prepared" and lists several things he should have done to prevent the situation from reaching a critical stage (51). Error creates a *différend* in medicine, not because it doesn't happen or happens rarely, but because medicine does not have a public idiom in which to discuss it. Indeed, error in medicine can only be communicated publicly in a legal language of malpractice or a media language of scandal: within these idioms, however, doctors and their mistakes in a sense still remain aberrant to medicine. Good doctors don't make mistakes, only "bad doctors" do, and these bad ones then become, appropriately it is usually believed, the subjects of malpractice suits and media scandals. Ironically, this is revealed even by the story of Gawande's own error: Would he have been able or willing to tell this story if it had not ended well?[7]

Because the idioms of malpractice and scandal don't confront error as an inevitable part of medicine, but as always already outside of its proper practices and procedures, Gawande explores the possibility of inventing a new idiom, or reworking an old idiom, for addressing error in medicine. He considers the idiom of the weekly Morbidity and Mortality (M & M) presentation that most hospitals conduct as "a cultural ritual that inculcates in surgeons a 'correct' view of mistakes" (61–62). Gawande notes that the M & M does at least acknowledge "that mistakes are an inevitable part of medicine," but that its ethos also, paradoxically, "reinforces the very American idea that error is intolerable" (62).[8] The M & M presents avoiding error as largely a matter of individual will, which doesn't dislodge what Susan Wendell calls the myth of control that she locates at the center of Western medicine. Gawande suggests if we understand error as a structural not an individual problem then we might develop a way of addressing the problem more effectively. Interestingly, it is an industrial management idiom on errors that Gawande thinks might help medicine counter errors more effectively in its practice. Gawande cites the work of British psychologist James Reason, whose book *Human Error* (1990) argues, according to Gawande, that "disasters do not simply occur; they evolve" (63). To understand error in medicine, we have to extend our investigation beyond the moment of the error in time and space: we have to trace back from an error in order to understand how it came to happen as the result of numerous small decisions made over time. As Gawande writes, "It isn't enough to ask what a clinician could or should have done differently so that he and others may learn for next time. The doctor is often only the final actor in a chain of events that set him or her up to fail. Error experts, therefore, believe that it's the process, not the individuals in it, that requires closer examination and correction" (64). Besides seeking new methods to reduce error in medicine, one of the secondary effects of this approach, which Gawande doesn't mention, is that it challenges the myth of control in medicine by de-heroicizing doctors. In the industrial approach, doctors are only another cog in a system of care that can either work efficiently or break down.

Gawande is clearly drawn to the possibility of transforming medicine through an encounter with systems analysis[9] and its idioms, but he also recognizes that, although some specific practices in medicine might benefit from this approach, medicine as a whole cannot simply be reduced to data for systems analysts to analyze. The point, it seems to me, is to understand that medicine is a multiple object, comprised of countless practices, many of which can and should be routinized and many of which can or should not be. Gawande suggests that a potential key to unlocking the problem of medical errors is "a technique called 'critical incident analysis,'" borrowed from the aviation industry and already utilized quite successfully to reduce errors in anesthesiology (66). Gawande explains that, "[t]he technique is built around carefully conducted interviews, designed to capture as much detail as possible about dangerous incidents; how specific accidents evolved and what factors contributed to them. This information is then used to look for patterns among different cases" (66). I suggest that a critical analysis of medical practices in general is essential to bring into being more effective accountings of what medicine gets right and what it gets wrong, which will lead to more effective practices of medicine. In order to successfully analyze its practices, medicine will have to open itself to researchers who study practices, and, as Gawande is well aware, doctors' own accounts of what they do and why will have to be open and honest.[10] Gawande's industrial idiom might provide new methods for medicine to confront some of its *différends*, but this approach still isn't complex enough.

Suffering

If error is one *différend* in medicine, suffering is another. The illness narratives I have analyzed in *Treatments* have all been attempts to negotiate suffering, and in this sense I read them as presenting affective histories. Gawande also negotiates suffering in *Complications*, and he emphasizes again how medicine often gets it wrong when it encounters suffering, because it doesn't have an idiom to communicate patients' experiences of suffering nor its own uncertainty when confronted with such suffering. Gawande explores suffering through two feelings: pain and nausea. His chapter on pain is called

"The Pain Perplex," and "perplex" is exactly the right word to use to describe medicine's encounter with pain. According to the *OED*, to perplex is: "To fill (a person) with uncertainty as to the nature or treatment of a thing by reason of its involved or intricate character; to trouble with doubt; to distract, confuse, bewilder, puzzle"; "[t]o torment, trouble, vex, plague"; "[t]o render (a thing) intricate or complicated in character and hence difficult to understand or deal with; to make (a thing) doubtful or uncertain through intricacy; to complicate, confuse, muddle"; and "[t]o bring into an intricately involved physical condition; to cause to become tangled; to entangle, intertwine; to intermingle." How can medicine communicate the perplexing experience of pain? How can medicine best treat pain? How can it create treatments that reduce rather than add to a person's experience of pain?

Gawande admits that for most doctors "chronic-pain patients are a source of frustration and annoyance: presenting a malady we can neither explain nor alleviate, they shake our claims to competence and authority" (118). Doctors confronted with their patients' pain are perplexed; they are fundamentally uncertain about what to do or say about this problem. Outright dismissal or more subtle avoidance strategies are quite common. Often these tactics are defended as necessary so that doctors can do their jobs without being overburdened by emotional strain.[11]

In his chapter "The Pain Perplex," Gawande discusses the changing theories of what causes pain, and essentially comes to the conclusion that pain is a biopsychosocial phenomenon, as Kleinman and others have argued (Kleinman 1988; Kleinman, Das, and Lock 1997). This is less interesting to me than his discussion in his chapter on the equally perplexing (and related) problem of nausea. There he describes the experience of Amy Fitzpatrick, a management executive who becomes incapacitated with nausea during pregnancy. "What is nausea," Gawande asks, "this strange and awful beast? The subject gets little attention in medical school, and yet, after pain, nausea is the most frequent complaint for which people consult physicians" (133). One thing that Gawande knows nausea is is "aversive," and he explains that this aversive feeling is remembered long after the nausea itself subsides. So, the object that is perceived to have caused the nausea—in Fitzpatrick's case pregnancy—continues to elicit somatic memories of the nausea whenever the object is encountered again or recalled.

Gawande's explication of the effects of nausea relates to Scarry's phenomenological description of pain, which elucidates the ways in which torture and war are structured such that they are capable of "unmaking the world" of an individual. The key term in Scarry's analysis is "structure": the unmaking of the world that happens through the experience of pain is not a by-product of, but structures, torture and war. According to Scarry, "Pain is a pure physical experience of negation, an immediate sensory rendering of 'against,' of something being against one, and of something one must be against. Even though it occurs within oneself, it is at once identified as 'not oneself,' 'not me,' as something so alien that it must right now be gotten rid of" (1985, 52). As Scarry sees it, this uncanny sense that one's embodied self has become alien is similar in illness and torture, but only in torture is this "internal physical experience . . . accompanied by its external political equivalent, the presence in the space outside the body of a self-proclaimed 'enemy,' someone who in becoming the enemy becomes the human embodiment of aversiveness" (52). Although I think Scarry is absolutely right to establish the specificity of the experience of pain in torture and to distinguish it from the experience of pain in illness, nonetheless, it can hardly be said that in illness the subjective characteristics of pain are *never* objectified or externalized. Ironically, this is precisely what medicine attempts to do, if usually in a benign, not averse manner, and often does so, as many of the illness narratives I have taken up in *Treatments* attest to, in such a way that it contributes to the patient's distress, and to the unmaking of the patient's world, because it fails to take seriously something it can't see.

Gawande explains that in general "our medical arsenal against nausea and vomiting is still fairly primitive" (141). Why this should be the case is one of the key questions Barbara Ehrenreich asks in "Welcome to Cancerland," and for her it is a political question. In the particular case of Amy Fitzpatrick's pregnancy sickness, Gawande explains that the situation had "moved beyond the doctors' control and worse, their understanding" (139). What do we do when we are in a situation beyond our control? In this moment a *différend* emerges: on the one side of the gap is Fitzpatrick's unremitting nausea and the suffering it causes her, on the other, her doctors' feelings of vulnerability because they lack the tools to effectively treat her nausea and address her suffering. The *différend* is the aversiveness Scarry describes: Fitzpatrick's nausea makes her experience her own body as averse,[12] and her doctors experience her experience of nausea and their lack of tools to control it as averse. Can we find an idiom to link these two experiences of aversiveness? Gawande suggests a new route: palliative medicine, or the "scientific study of suffering" (142). Like systems analysis to address the *différend* of errors in medicine, this route isn't particularly radical, except in its insistence that suffering is something that science must address. One technique the palliative specialists believe might help communicate a patient's suffering to her doctors, and therefore contribute to its alleviation, is establishing pain as a fifth vital sign; along with temperature, blood pressure, pulse, and respiratory rate, health care practitioners would monitor pain—"the level of discomfort a patient reports" (143). Palliative specialists are interested in creating tools— like the "McGill Pain Questionnaire," discussed in *The Body in Pain* as an important "diagnostic tool that enables patients to articulate the individual character of their pain with greater precision than was previously possible" (Scarry 1985, 7)—that work to externalize suffering so that it can be communicated, treated more effectively, and even anticipated before it becomes aversive.

In the end, nothing helps alleviate Fitzpatrick's nausea; one day it just stops. She tells Gawande that the doctors "she liked best were the few who admitted they didn't know how to explain her nausea or what to do about it. They would say that they had never seen anything like her case, and she could tell that they commiserated with her" (144).[13] Yet another tool, then, might be called "practicing at a loss,"[14] which would require that doctors give up the myth of control, acknowledge their failure to always have all the answers, and to not turn away from suffering. Practicing at a loss requires being with a patient not only as practitioner and expert, but also as witness.

Autonomy

The final *différend* I want to address is not one Gawande's analysis reveals in medicine, but one that appears in Gawande's analysis itself, of medical decision-making and the historical emergence of the concept of a patient's autonomy in this process. I don't disagree with Gawande's overall premise that decision-making is a negotiation between doctor and patient; this practice of negotiation between doctors and patients is a shift that I have sought to document as well. He is surely right that many patients believe doctors are best positioned to make decisions about medical treatments; after all, medical treatments are what doctors are trained to administer. So, I don't find it problematic that Gawande would suggest that we have made something of a fetish of a patient's autonomy at the expense of effective decision-making. What I do find problematic, however, is his interpretation of how the autonomous patient became the ideal within Western medical culture. In Gawande's account of the emergence of the autonomous patient, that emergence had nothing to do with patients advocating for themselves or with the politicization of patienthood, but with doctors and their changing view of what a good patient should be and do. Doctors, in Gawande's telling, decided patients should be autonomous.

Gawande's history goes like this: "Little more than a decade ago, doctors made the decisions; patients did what they were told" (210). And then this all changed. Gawande explains that, "[o]ne of the reasons for this dramatic shift in how decisions are made in medicine was a 1984 book, *The Silent World of Doctor and Patient*, by a Yale doctor and ethicist named Jay Katz" (210). Katz's book is a seminal work in bioethics, but to cite it as the one source of this dramatic shift worth noting is to reveal a *différend* in Gawande's own ethics of failure: let's call it politics, or feminist politics in particular. That patients should make their own decisions about their treatment emerged with the women's health movement of the 1970s, which influenced AIDS activism in the 1980s and breast cancer activism in the 1990s. Moreover, a critique of autonomy emerged with a feminist ethics of care, first articulated in Carol Gilligan's *In a Different Voice*, which was published in 1982, and which itself has been extensively critiqued and complicated since then.[15] Many feminist scholars have argued that selves and bodies in the world are not autonomous and sovereign, but always come into being in relation to others. More recently, some feminists have also emphasized vulnerability as a universal human experience. Donna Haraway writes in "The Biopolitics of Postmodern Bodies": "Life is a window of vulnerability. It seems a mistake to close it. The perfection of the fully defended, 'victorious' self is a chilling fantasy . . . whether located in the abstract spaces of national discourse, or in the equally abstract spaces of our interior bodies" (1991, 224; see also Price and Shildrick 2002). I mention what is missing from Gawande's account not because I believe feminist health activism and bioethics are the only routes to begin to address the *différends*—error and suffering—in medicine, but to indicate, as Lyotard would, the multiplicity of phrase regimens and the difficulty of making links between them. When Gawande turns from medicine to other idioms, he turns to systems analysis and palliative medicine. Where else might he turn? This question must remain open; it is a mistake to close it.

Rose's Diaporia

If the key terms in Gawande's work are fallibility, mystery, and uncertainty, the key terms for British philosopher Gillian Rose are uncertainty and contingency. In *Love's Work* (1995), she attempts to describe the experience of illness—her own illness and the illnesses of others—without resorting to the compensatory imaginary of a heroic cultural narrative. Rose's work is both a memoir and a philosophical essay; in fact, it might be called a philosophical memoir. Like *Complications*, *Love's Work* allows us to consider, as I have attempted to do throughout *Treatments*, what Foucault called the "fundamental structures of experience" that link medical thought, philosophical thought, and literature (1973, 198). Rose's linkings are different from Gawande's, partly because her routes begin from a different position than do his. Hers begin with philosophical thought and its phrases, which she then links, or tries to link, to medical thought and its phrases, which she encounters through her experience of conventional and alternative medicine as a patient, not a practitioner. Finally, she links these philosophical and medical phrases to an invented literary form she calls "colostomy ethnography." In many respects, Gawande's *Complications* and Rose's *Love's Work* are incommensurable texts, but by bringing them together here I want to encourage my readers in the endless task of "bear[ing] witness to differends by finding idioms for them" (Lyotard 1988, 13). What do we turn away from and what do we turn toward? This is a fundamental question for literature, philosophy, and medicine. In *Complications*, a doctor writes in the hopes of finding practical methods to help medicine treat patients more effectively, which means treating their suffering as much as their physical symptoms. In *Love's Work*, a philosopher who is dying writes to get to a realm beyond medicine. In *Treatments*, I write about *Complications* and *Love's Work* together in the hope that drawing their multiple idioms together will suggest a method medicine might make use of to invent new ways of thinking what it does.

City of Death

Love's Work opens with Rose's return visit to New York City to see her friend and former lover, Jim. New York is where Rose discovered her life's work, continental philosophy and critical theory, and Jim was integral in that discovery. Her first visit to New York in August 1970 was just after she'd graduated from Oxford. Oxford, for Rose, had been an educational experience that wasn't particularly educational and couldn't have ended soon enough. Her real education, she tells her reader, didn't begin until New York, a trip that was supposed to be for a mere three weeks, but became a year-long sojourn. Yet, her return to New York in 1991 affords her a new view of the city and a different sort of education. And, in fact, this return begins with misrecognition, and the possibility that her life has diverged from Jim's. At the airport, Rose explains, she initially goes up "to someone who looked like a caricature of Jim as I remembered him in good health" (1995, 4). She literally *doesn't see*, or perhaps *doesn't want to see*, the Jim who has come to meet her, transformed as he is by HIV/AIDS. Rose writes:

> My formerly laconic and witty friend had become loquacious, needy, addressing with urgent familiarity everyone we chanced to have dealings with over the next few days—taxi-drivers, bell-boys, waiters. And when he wasn't holding forth to those nearest to him, he issued a continuous, low, moaning sound, a piteous cradling for the inner, wounded being that, strangely, had surrendered to the publicity of the city streets. (4)

Rose is disconcerted by the ways in which Jim's private moaning moves into the public; it's as if his most private pain leaks out into the world, and this leakage is something he can no longer control, nor is even aware of. Not unlike Gawande's dismissive doctors when confronted with the chronic pain of their patients, Rose feels an aversion to Jim's public moaning; it reveals a loss of control that she lacks the tools to address, except after the fact in her own illness narrative.

It seems everyone Rose meets in New York this time is "extraordinarily afflicted" (7); in fact, the city itself is extraordinarily afflicted, and she calls it a "city of death."[16] What Rose doesn't say, because she doesn't know at the time, is that she will be—indeed, perhaps, already is—one of the extraordinarily afflicted; and her affliction—ovarian cancer, the "silent" cancer—will only be discovered, like 80 percent of its cases, in an advanced state, having already metastisized to her colon (83). Her own inner, wounded being remains cradled within, and does not—yet—surrender to the publicity of the city streets. The reader doesn't learn about Rose's illness until halfway through *Love's Work*, and only after she has wondered how her reader would respond if she were to say she has AIDS—"full-blown AIDS"—and that she has been "ill during most of the course of what [she has] related" (76). What, Rose wonders, does the revelation of illness (and, presumably, imminent death) do to the reader, who she interpellates here as "you"? Answering her own query, Rose insists:

> I would lose you. I would lose you to knowledge, to fear and to metaphor. Such a revelation would result in the sacrifice of the alchemy of my art, of artistic "control" over the setting as well as the content of your imagination. A double sacrifice of my elocution: to the unspeakable (death) and to the overspoken (AIDS). (76–77)

And yet, Rose recognizes the uses of losing us, and of losing control over the setting as well as the content of our imagination. The possibility of losing her reader and of losing control over her story are not reasons for not writing; rather, the possibility of such loss—the inevitability of such loss—is reason itself for writing. No, she doesn't have AIDS; rather, she has something else that is "full-blown," and that is not overspoken,[17] but barely spoken, or, perhaps, like the disease itself until it is too late, silent. Not to write because she cannot control this story is to "die deadly" (77). And yet, she explains to her reader and, in many ways, to herself as well, "by this work, I may die forward into the intensified agon of living" (77). Writing is not a means for overcoming the difficulty of dying, but for engaging with the agon of living: dying forward is not simply leaving a record of one's heroism in the face of one's death, it is grappling with one's inevitable failures in one's reckoning with life, despite knowing that the outcome of the grappling, and the story of the grappling, is unknown.

A Realm beyond Medicine

As we've seen throughout *Treatments*, many critiques of modern medicine, especially feminist ones, have noted the ways in which the patient is objectified and silenced within the isolating spaces and hierarchical relationships created by the anatomo-clinical method. In ethical terms, the doctor-patient relationship is often characterized not by the modernist ethics of beneficence and nonmalfeasance, but by paternalism at best and punitiveness at worst. Rose experiences conventional medicine as a "carnival of communication" (97) when two consulting surgeons—Mr. Wong, the gynecologist, and Mr. Bates, the bowel specialist—report vastly different prognoses after a second operation on her bowel. Mr. Wong, on the one hand, reports that Rose's cancer has spread considerably, and that her prognosis is "guarded" (99). Mr. Bates, on the other hand, assures Rose that she is "living in symbiosis with the disease," and should "[g]o away and continue to do so" (100). Their "utterly discrepant opinions" (100), while perhaps not averse in themselves, are made so by the fact that what Rose's consultants are most concerned with is not her well-being, but whose position is taken as authoritative in relation to Rose's disease. Mr. Wong refuses to talk to his colleague; as he tells Rose, "I will not change my position. This is my cancer" (101). Rose, in a panic, must plead, cajole, beg, flatter, and inveigle Mr. Wong to talk to Mr. Bates (101). In this absurd scenario, Rose's body becomes the ground upon which disputes over power are fought. There is no person with cancer in this scenario, there is only a cancer, either Mr. Wong's or Mr. Bates's, depending on who is speaking and, in speaking, has the power either to cure or condemn it (103).

Rose eventually realizes that she is "already in a realm beyond medicine," and that she and medicine "do not have enough command of each other's language for the exchange to be fruitful" (102). Rose discovers that she and medicine are mute to each other, and this muteness is another case of a *différend* between two parties, which, Lyotard explains, "takes place when the 'regulation' of the conflict that opposes them is done in the idiom of one of the parties while the wrong suffered by the other is not signified in that idiom" (1988, 9). This scene reveals both *différends* my discussion of Gawande already identified: error and suffering. Because Mr. Wong and Mr. Bates disagree on her prognosis, at least one of them must be mistaken, but they have no way to overcome the incommensurability that arises if one or both of them is wrong. Both doctors are also unable to speak of the wrong suffered by Rose; they can only speak of their capacity to properly diagnose her disease. And yet, through the practice of writing (and reading) her memoir, this incompatibility between idioms becomes an opportunity for a "philosophical politics" to "bear witness to differends by finding idioms for them," (Lyotard 1988, xiii, 13). This philosophical politics emerges not from the success of heroic medicine to cure Rose's cancer, but from her realization that her illness is in a realm beyond medicine, and that to speak of it requires a new idiom. She must turn away from medicine and the false hope of control it might give her.

Like Lyotard, who is concerned with the state signaled by the feeling "[o]ne cannot find the words" (13), Scarry is concerned throughout *The Body in Pain* with the failure of language in the face of pain. Her work is often read pessimistically; according to this reading, pain is incommunicable because, as Scarry notes, "[t]o have pain is to have *certainty*; to hear about pain is to have *doubt*" (Scarry 1985, 13; emphasis in original). Thus, she writes, "To witness the moment when pain causes a reversion to the pre-language of cries and groans is to witness the destruction of language" (6). However, she does not stop at the destruction of language. Rather, she insists that "conversely, to be present when a person moves up out of that pre-language and projects the facts of sentience into speech is almost to have been permitted to be present at the birth of language itself" (6). Scarry, like Lyotard, Gawande, and Rose, then, is concerned not only with the "difficulty of expressing physical pain," but also with the "nature of human creation," with the ways in which, through "acts of making, human beings become implicated in each other's sentience" (176).

Two acts of making in *Love's Work,* in which the process of making cannot be separated from the product that is made, are love and work, or, juxtaposing them as does Rose, *love's work*. "Love's work" is a *form of treatment*—an affective and effective practice of the self—that is, according to Rose, in opposition to the "iatrogenic materiality of medicine and to the screwtape overdose of spirituality of alternative healing" (1995, 77–78). "Love's work" is both a theory and a method for negotiating loss; it is a theory/method that conveys "the impasses, the limitations and cruelties, equally, of alternative healing and conventional medicine" and "insinuate[s] *démarches* [steps, proceedings, treatments] of healing that have not been imagined in either canon" (77). These not-yet-imagined *démarches* of healing are an opening to an ethical moment that requires both love and work, and their juxtaposition. Rose allows us to begin to see how the affective and the effective are implicated in each other. Love and work come together, and it is the experience of illness (the experience of unmaking that occurs in illness) that brings them together.

Colostomy Ethnography

In *The Body in Pain*, Scarry is interested in both the making and unmaking of the world, and the ways in which analyzing the processes and structures of unmaking might give us insight into the processes and structures of making, just as Gawande sought to analyze the evolutionary structure of medical errors. In *Love's Work*, Rose moves from a chapter on love to a chapter on illness. In her text, then, love and illness are juxtaposed not as opposites, but as two scenes of loss (of self, of certainty, of control) in which the embodied self is always already vulnerable. The work doesn't heroically disregard vulnerability and uncertainty, it acknowledges and negotiates with it. She opens herself up to her readers, but this opening reveals not just life in its medical sense as incommensurable with death, but life in its "meaningful sense . . . [as] inclusive of death" (79). The literal and figurative opening that Rose reveals is the opening of a colostomy, a procedure she must undergo because her cancer has spread to her colon. Of this decidedly uncanny treatment, Rose notes:

> Nowhere in the endless romance of world literature (my experience is, needless to say, limited) have I come across an account of living with a colostomy. Since the first colostomy was performed in this country in 1797, the first paper on the subject published in 1805, and colostomies have been routine medical practice since the second half of the nineteenth century, this is more than enough time for lyric and lament. (93)

In fact, it is anti-lyric and anti-lament that Rose practices: "Let me make myself clear," she writes, "the colostomy—*stoma* meaning 'opening'—is a surrogate rectum and anus" (93). Linking what she calls "colostomy ethnography" (94) with Holocaust ethnography,[18] Rose intends to speak of shit, to re-site bodily function that usually remains hidden out into the world, to exchange "discretion for an anterior cloaca and incontinence" (95).

With her "colostomy ethnography," Rose attempts to show the ways that illness disrupts the stability of binary oppositions, including inside/outside and public/private. Illness might be said to transform that which keeps these positions separate, to make the boundary between them vulnerable and porous. Thus, by speaking and writing indiscreetly of her body and its failures, Rose creates an opening that leads both ways: inside out and outside in. Because she is "already in a realm beyond medicine," and because "the language of medicine is irreconcilable with her own language," who she is exceeds what medicine might say about her. But it is not only conventional medicine that fails to provide Rose with a grammar and syntax with which to describe "a new bodily function" and the difference it does and doesn't make (95). Rose sees alternative healing practices as offering little more than compensation for friends and family who "field the crisis of their own mortality brought on by my illness by serving hard and fast at me the literature and liquids of alternative healing" (104). As Rose sees it, alternative healing practices provide a discourse for others (not the person who is ill) to take up in the face of the uncertainties that illness reveals. They may provide consolation, but rarely to those who are ill and already beyond consolation. According to Rose, moreover, "The injunction, which pervades the literature of alternative healing, to become 'exceptional' (Bernie Siegal), or 'edgeless' (Stephen Levine), to assume unconditional love, is poor psychology, worse theology and no notion of justice at all" (104). In contrast to the alternative healing discourses, the love, psychology, theology, and justice that Rose practices (and the narrative treatments she offers) is conditional, meaning, for Rose, *not absolute but speculative*. Alternative healing, in contrast, provides a "counsel of despair which would keep the mind out of hell" (105). In other words, it attempts to speak only in terms of a myth of control and heroic success in overcoming difficulty, and of the possibility of unconditional health and unconditional love. Failure and conditionality are radically excluded from this heroic mode. Alternatively, Rose proposes an ethics based on her book's epigraph: "Keep your mind in hell, and despair not."[19] She is concerned with an ethics that acknowledges that we are all "vulnerable, woundable, around the bounds" and that "conditionality is the only unconditionality of human love" (106). We must risk relation; we must seek the agon as well as the repose of living.

What Rose articulates in *Love's Work* is an ethics of failure, which is not at all the same as, or even compatible with, Gawande's ethics of failure, but productive of new idioms nonetheless. In fact, the term itself comes from Gillian Rose's sister, literary theorist Jacqueline Rose. In a lecture entitled "Why War?" Jacqueline Rose discusses the problem of war in general, and a dispute within the British Psycho-Analytical Society in 1943–44 between Melanie Klein, her supporters, and her critics in particular. Jacqueline Rose asserts, following British psychoanalyst D. W. Winnicott, that "[k]nowledge will be possible only if we are willing to suspend the final purpose and ends of knowledge in advance" (1993, 36–37). If we are to avoid going to war, Rose writes, we must, "[h]ang on to failure" (37). Knowledge itself requires a certain suspension of belief in the possibility of total knowledge. It requires a certain willingness to trust in that which cannot be known, or can only be known conditionally. Gillian Rose, similarly, understands ethics as the *diaporia*, or, "being at a loss yet exploring various routes, different ways towards the good enough justice, which recognises the intrinsic and contingent limitations in its exercise" (1995, 124). An ethics, for Rose, will never come from dissolving "the difficulty of living, of love, of self and other, of the other in the self," it will come from being at a loss yet exploring various routes (105). These routes cannot be mapped in advance, nor will they necessarily be mapped in any conclusive, totalizing way even as we travel along them. If what I take from Gawande is a notion of practicing medicine at a loss, I take from Rose a notion of practicing ethics at a loss.

Though Gawande seemingly unconsciously avoids the idioms of politicization that I have discussed throughout *Treatments*, Rose explicitly opposes feminism (and postmodernism)[20] in *Love's Work*. For Rose, both feminists and postmodern philosophers are "in deadly, unironic earnest" (125), and they fail to recognize "the way that play (fairy stories, terrifying films) teaches the difference between fantasy and actuality" (126). Rose admits that "feminism never offered me any help," and this doesn't seem surprising considering that she also claims feminism "fails to address the power of women as well as their powerlessness" (140). Clearly, Rose has a rather limited view of feminism; her view is a caricature of even radical second-wave feminist efforts to expose everyday sexism and violence against women. Her portrait of feminism might have been more generous had she considered the positions of feminists like her sister, who sees "radical self-questioning . . . not as a block or a ban to political life, but as a necessary part of its procedures" (1993, 232). Jacqueline Rose understands the feminist project as "not exactly utopian but more a bid for the future" (232), which seems to be what has motivated feminist health activism and AIDS activism. This bid for a future is both literal and figurative: think of Audre Lorde's imagined community of one-breasted women sharing knowledge with each other and Paul Monette's willingness to have "insomnia for everyone now" because he can sleep when he's dead.

Gillian Rose, ultimately, maintains a "skeptical faith, shaky but persistent, in critical reason" (139), and she believes such a skeptical faith in critical reason is not to be found in feminism or postmodernism, nor, for that matter, in a postmodern feminism. In this respect, Rose might seem to come down on the modern side of the *différend* that Lyotard describes in *The Postmodern Condition*. The two modes Lyotard describes "testify to a difference *(un différend)* on which the fate of thought depends and will depend for a long time, between regret and assay" (1984, 80). The difference, for Lyotard, is an aesthetic one, between an aesthetics that regrets what has been lost and aesthetics that assays, that is, "searches for new presentations" (81), and an ethical one, between a model of legitimation that emphasizes consensus and the maximization of performance and one that emphasizes pragmatic paradoxes and the production of the unknown. Although Rose voices a persistent faith in critical reason that bears some similarities with Croce's position in "Discussing the Undiscussable," her faith is not a nostalgic, sentimental (romantic?) one, meant to offer solace and simple pleasure by keeping experience, politics, and art in separate domains, but a faith that, paradoxically, requires risk and speculation as she explores the boundaries between fantasy and actuality: "I will stay in the fray, in the revel of ideas and risk; learning, failing, wooing, grieving, trusting, working, reposing—in this sin of language and lips" (144). Although Rose's readings of the theories and practices of feminism and postmodernism differ from those I offer in *Treatments,* I want to heed her advice even as I critique her readings: to stay in the fray and revel of ideas and risk, and of ideas that take risks rather than offer comfort.

Doctors and patients are brought together by illness and a *différend* expands between them. We must risk failure, risk relation; because by taking such risks, we open up the possibility of new routes, new treatments: in and between art, medicine, philosophy, and politics. There are no shortcuts, however; on that Rose would agree with Iris Murdoch. And, it is not enough simply to survive; on that Rose would agree with Audre Lorde and Paul Monette. "To survive—to live again," Rose writes in *Mourning Becomes the Law* (1996),[21] "demands a new tale: a new prayer to be found, a new polity to be formed" (100). The tale is one that includes prayer and polity, poetics and politics, affective and effective history. Telling this new tale requires work: the work of mourning, the work of politics, the work of language, the work of love. An ethics of failure takes failure not as an ending, but as a beginning.

Acknowledgments

Treatments is about the multiple arts of being ill and doing illness as presented in illness narratives. Illness changes our relationships to ourselves, to others, and to the world around us. While illness often does isolate people who are ill, it also frequently motivates them to create new social arrangements, as well as new intellectual and political communities. The process of writing *Treatments* was similar: although much of the research and writing was done on my own, this process has also brought me in contact with people and ideas that have changed me and the way I understand the practice of doing treatments.

One of the things that this book explores is what Dori Laub has called "the discovery of knowledge." When and where does that "discovery of knowledge" begin? I can only begin to trace a genealogy that has led to the emergence of *Treatments*. While studying for a master's degree in women's studies at Lancaster University in the United Kingdom, I first took courses on feminist approaches to health. Margrit Shildrick was my teacher, and she remains a mentor to me; her imprint is all over this work. At Lancaster I was also fortunate to work with Sara Ahmed, Beverley Skeggs, and Clare Whatling. At Emory University, Angelika Bammer and Walter Kalaidjian influenced this project. They encouraged me every step of the way, and some of the ideas discussed here originated in classes I took with them. I'm also grateful for what I learned from Pam Hall, Tom Flynn, Robert Paul, and David Carr.

At Stony Brook University, many of my colleagues have been much more than simply collegial. Mary Rawlinson brought me to Stony Brook and has been a steadfast supporter of my work ever since. She asked Nancy Tomes to mentor me when I arrived, and I feel truly blessed by that decision and grateful for Nancy's ongoing support. I am thankful for the opportunity to know Kelly Oliver and Benigno Trigo, although they are no longer at Stony Brook. I benefited greatly from discussions with Kelly on psychoanalytic practices of witnessing and with Beni on literature, medicine, and imperialism; both have taken the time to read my work and comment on it. Through our mutual interest in the emergent field of medical humanities and through a friendship that goes beyond that shared interest, Catherine Belling and I have managed to bridge the literal and figurative divide that separates the medical school from the humanities and social sciences at Stony Brook. Christa Erickson helped me to begin to think about the potentially creative relationship between medicine and visual art. And many others at Stony Brook have supported this work by providing an invigorating intellectual community, including Bonnie Gordon, Ann Kaplan, Manuel Lerdau, Shirley Lim, Sara Lipton, Celia Marshik, and Martha Mockus.

Many people have read pieces of this book and made helpful suggestions to improve my argument or writing; they include Jonathan Ablard, Rita Charon, Thomas Couser, Rosemarie Garland-Thomson, Aran MacKinnon, Jonathan Metzl, Annemarie Mol, Suzanne Poirier, Alan Radley, and Martha Stoddard-Holmes. Richard Morrison at the University of Minnesota Press has been a stalwart supporter from the beginning and has guided me through the editorial process with skill and good humor. The anonymous readers for the University of Minnesota Press were both incisive and generous in their comments. They offered many practical suggestions to make *Treatments* a better work, and I am most grateful for that. Thanks also to Laura Westlund and Mary Lou Kenney, whose close attention to the manuscript has made my writing sharper.

My family has always encouraged me to be curious about the world, and this is the beginning of the critical consciousness that informs my treatments here. I thank my mother, Francyne Diedrich, for her interest in this project from its earliest stages and for her strong views on the practice of memoir writing. Many thanks as well to my father and stepmother, Richard and Linda Diedrich, and my sisters, Dawn Diedrich and Andrea Diedrich Kumar. My nephews and niece, Jack Boyette and Nikhil and Sona Kumar, have grown into smart and funny individuals with independent minds during the many years I have been working on this book.

Two people in particular have shaped this project and me the most. Cindy Patton was the advisor for the dissertation that was the foundation for this book, but she has been and continues to be so much more than an advisor to me. She pushed me to think about the cultural work illness narratives do and challenged me to defend my positions in a bolder and less suggestive manner. Most important, she is a shining example of how to engage ethically with ideas, other people, and the world. I will be forever trying, and failing, to live up to her example.

Finally, Victoria Hesford has had to live with this book through all of its ups and downs. She has read every word, in some cases several times, and has frequently got me out of a bind or into new territory with a well-timed critical intervention. She has encouraged me to be more adventurous in my thinking and has constantly reminded me that thinking can be a radical act. Most of all, she has taught me the dangers of taking myself too seriously, and the pleasures in not doing so. That lesson seems central to the treatments here.

Notes

Introduction

1. Throughout *Treatments*, I refer to the autobiographical texts that provide much of the empirical data for this study as "illness narratives," a term that I take from psychiatrist Arthur Kleinman's *Illness Narratives: Suffering, Healing, and the Human Condition* (1988). Although I situate the rise of illness narratives within a more general "age of memoirs" in the West, not all of the texts I take up in the following chapters fit a rigorous definition of "memoir," which is why the broader term "illness narrative" is more appropriate for my study: for example, in chapter 1, *Madonna Swan* might be better described as an oral history, in chapter 2, Sontag's *Illness as Metaphor* and *AIDS and Its Metaphors* are first and foremost essays on literature and culture, and in chapter 3, Ruth Picardie's *Before I Say Goodbye* combines newspaper columns, diary entries, email correspondence, and so on. For me, the term "illness narrative" is an umbrella term under which these sometimes disparate forms might fit.

Kleinman's important work describes the ways in which illness takes on personal and cultural meanings, as well as the ways that illness and its meanings are given narrative form. He focuses, in particular, on how narrative helps to bring order and give coherence to the experience of illness (49). In his work, Kleinman does not look at published accounts of illness experiences; rather, he employs an ethnographic methodology and offers ethnographic case studies that describe the multiple meanings that surround illness, especially chronic illness. Kleinman utilizes ethnography to gather data for his research on the meanings of illness and also recommends ethnography as a form of clinical and therapeutic practice. In other words, he encourages doctors to become participant observers within their patients' particular social milieus. For my purposes, Kleinman's methodology is most useful for the challenge it offers for medical practitioners to the biomedical model of care, and I return to his text in my chapter on doctors' narratives and the doctor-patient relationship.

2. For an analysis of the cultural work of early representations of AIDS in the mainstream media, see Treichler 1999, particularly chapters 4 and 6.

3. In his latest book, *The Renewal of Generosity*, Arthur W. Frank notes at the outset that, "the foremost task of responding to illness and disability is not devising new treatments, though I'm grateful this work will proceed" (2004, 1). In my work, I want to complicate what it means to devise treatments, moving beyond the strictly medical and into other realms, including the literary, philosophical, cultural, and political.

4. I take this term from Adam Phillips (2000, 116).

5. As a polemic, "Discussing the Undiscussable" has provoked much discussion; there was a vociferous response in newspapers and magazines immediately after its publication (see, for example, Leo 1995, Gilbey 1995, and Rich 1995), and both Elizabeth Spelman and G. Thomas Couser discuss Croce's essay at length in books published in 1997. Furthermore, when Bill T. Jones's "Still/Here" is performed in a new place, the story of Croce's critique is often, ironically, a part of the publicity surrounding the show (see, for example, Gilbert 1998 and Mackrell 1995).

6. Two years prior to the publication of Croce's piece, *Newsweek* had a cover story called "AIDS and the Arts: A Lost Generation." In that article, Bill T. Jones admits that the death of his partner, Arnie Zane, of AIDS in 1988 "was a kick in the pants. It forced us to ask the question, what is the relation between life, creativity and death?" (Ansen 1993, 20).

7. Thanks to Victoria Hesford for suggesting this contrast to me.

8. Kushner's was one of several letters responding to Croce's piece published in the January 30, 1995, issue of the *New Yorker*. Camille Paglia, bell hooks, Hilton Kramer, Midge Dexter, and others also sent letters commenting on the piece.

9. I'm not the only reader who found this sentence particularly troubling; see also Brennan 1995 and Mars-Jones 1995. Mars-Jones deconstructs this sentence thoroughly, and parenthetically remarks, "If you do see yourself as reviewing people rather than what they make, then a stroppy gay black man with HIV may well be a tough assignment" (24). Several commentators allege that Croce's article is little more than a veiled personal attack against Jones. For example, in a letter to the *New Yorker*, bell hooks contends that Croce uses the public forum of the *New Yorker* in an attempt "to 'win' whatever personal war she has chosen to wage against Bill T. Jones" (New Yorker 1995, 10). Relatedly, Elizabeth Spelman, in an impressive reading of the piece in relation to the history of moral philosophy, remarks, "The fierce critic of 'victim art' has, not to put too fine a point on it, claimed victim status for herself" (1997, 145).

10. Thanks to Richard Morrison for providing me with a copy of the program.

11. And, as I will take up below in my discussion of the category "experience," what would this look like anyway?

12. The qualifiers "always" in the title and "necessarily" in the quotation are significant, and make his piece less polemical than Croce's.

13. Here Clarke follows Olney's foundational insight in autobiographical studies that, "[a]utobiography is both the simplest of literary enterprises and the commonest" (1980, 3). In the same essay, Olney also notes that, "autobiography is . . . the least

'literary' kind of writing, practiced by people who would neither imagine nor admit that they were 'writers'" (4).

14. He names, as an example of an aesthetically good memoir, Mary Karr's *Liar's Club* (1995).

15. The novel, of course, was criticized at the time of its emergence as less than true art. The prominence of "lady" novelists in the nineteenth century as a sign of the novel's questionable pedigree is similar to the argument made against the memoir a century later. Thanks to Victoria Hesford for reminding me of this.

16. See Driver 1994 (115) for a useful discussion of Foucault's idea of history as disconcerting.

17. For a discussion of genealogy as a useful method for disability studies, see Diedrich 2005.

18. In her book *Volatile Bodies* (1994), in which she outlines the need for a "corporeal feminism," Grosz takes up philosophies that are either directed "inside out" (for example, Merleau-Pontian phenomenology, Lacanian psychoanalysis, and neurophysiology) or directed "outside in" (for example, the philosophies of Nietzsche, Foucault, Deleuze, and Lingis). According to Grosz: "Whereas psychoanalysis and phenomenology focus on the body as it is experienced, rendered meaningful, enmeshed in systems of significations, Nietzsche, Foucault, Deleuze, Lingis, and others . . . focus on the body as a social object, as a text to be marked, traced, written upon by various regimes of institutional (discursive and nondiscursive) power, as a series of linkages (or possibly activities) which form superficial or provisional connections with other objects and processes, and as a receptive surface on which the body's boundaries and various parts or zones are constituted, always in conjunction and through linkages with other surfaces and planes" (116). As Grosz makes clear in this description, the focus in both of these philosophical modes is on the body, or, perhaps, the self as embodied.

19. My conception of the emergence of this figure came out of fruitful discussions with Victoria Hesford on her work on the emergence of the feminist-as-lesbian figure in the second-wave women's liberation movement (Hesford 2001).

20. For an informative reading of the case history and medical chart as narrative, see Epstein 1995, and for an important analysis of the "narrative structure of medical knowledge," see Hunter 1991. There is also a growing body of work on narrative ethics in medicine; see, for example, Chambers 1999 and Charon and Montello 2002.

21. For example, even Margery Kempe's autobiography, dictated in the fifteenth century and described by many as the first autobiography in English, might be considered an illness narrative. Thanks to Angelika Bammer for bringing this to my attention. By describing the emergence of a particular genre and figure, I do not mean to say that there are no examples of such writing prior to 1980. What happens around 1980, however, is a more general politicization around the experience of patienthood, as well as the widespread sense that the experience of patienthood can be transformed in and through writing.

22. In her attempt to create an archive of lesbian life and trauma, Ann Cvetkovich acknowledges the "difficulty of gathering even a history of the present and the challenge, even for experimental ethnography, of documenting emotional life" (2003, 10). She explains that we must "forge methodologies for the documentation and

examination of the structures of affect that constitute cultural experience and serve as the foundation for public cultures" (11). I too want to document structures of affect that constitute cultural experience, which is one reason why I take up a popular genre and the responses to it.

23. My conception of *doing* illness and *doing* treatments is influenced by Annemarie Mol's work, in particular in *The Body Multiple* (2002), on the multiple ways researchers, clincicians, and patients do atherosclerosis.

1. Patients and Biopower

1. There is an emerging literary critical field on illness narratives, which has influenced my approach in *Treatments*. Along with Frank's *Wounded Storyteller* (1995) and *Renewal of Generosity* (2004), some of the key works in this emergent field include: Gilman 1988; Hawkins 1993; Murphy and Poirier 1993; Couser 1997; Morris 1998; and Nelson 2001. There are of course also works that consider the depiction of the profession of medicine and/or the experience of illness in other literary genres: novels, short stories, drama, and poetry, and also on representations of suffering in visual art and film (see, for example, Elkins 1999 and Radley 2002). I have chosen to limit my study to memoirs about illness because this book is as much concerned with historically situating the rise of the memoir in late-twentieth-century Western culture and with problematizing the confessional mode as with explicating the themes of the narratives themselves. For examples of work in the area of literature and medicine and the emergent interdiscipline of medical humanities, see Brody 1987, Clarke and Aycock 1990, and essays in the journals *Literature and Medicine* and *The Journal of Medical Humanities*.

2. As in *The Wounded Storyteller,* in his latest book, Frank invents a figure through which to explore the experience of illness; here the "Dialogical Stoic" replaces or expands on the role of the wounded storyteller. As its name suggests, the Dialogical Stoic juxtaposes the arts of existence of the stoic philosopher Marcus Aurelius and twentieth-century Russian literary critic Mikhail Bakhtin, who elaborated a literary critical method known as "dialogics." Frank realizes when he is recovering from a serious illness that he experienced illness in two modes, in isolation and with others, or perhaps more accurately, as two temporalities, the time spent alone cultivating an inner dialogue and the time spent with other people that revealed his dependence "on their imagination of [him]" (2004, 32). Frank's Dialogical Stoic is a recognition of the self as formed in relation to itself and to others. Both of these relations are constantly in flux, as the self is always changing in relation to itself and others (who are also always changing). I like Frank's Dialogical Stoic because it shows the way juxtaposing two seemingly incommensurable theories of the self might help us better understand the experience of illness. Moreover, Frank's two modes move inside out (Stoicism) and outside in (dialogism), both movements that I also understand as essential to understanding the experience of illness.

3. For the most part, Frank uses the terms "modern" and "postmodern" to differentiate between historical periods: the modern covers the period following the eighteenth-century Enlightenment in the West, up until and including the first half of the twentieth century, the period which saw the professionalization and institution-

alization of medicine in the West (see Starr 1982 and Foucault 1973), while the postmodern refers to the latter half of the twentieth century. Frank characterizes modern and postmodern medicine in such a way that medicine might be seen as a microcosm for more general theoretical and cultural shifts, exemplified by the Enlightenment and criticism of the Enlightenment. See also David Morris's delineation of illness within postmodernity in his book *Illness and Culture in the Postmodern Age* (1998).

4. Although *The Wounded Storyteller* is mostly a literary critical work, Frank does on occasion refer to his own experience of illness (which includes a heart attack in 1985 caused by a viral infection and testicular cancer shortly thereafter), and describes that experience as a personal and political impetus for his work. He has also written his own illness narrative entitled *At the Will of the Body* (1991).

5. See, for example, the extensive feminist critique of postmodernism, including Fraser and Nicholson 1990, Brodribb 1992, Mohanty 1994, and Benhabib 1995. Which isn't to say, of course, that all feminist theory is antipathetic to postmodernism. Many of the feminist work I take up in this book—including Braidotti 1994, Probyn 1993 and 1996, Singer 1993, and Grosz 1994—make use of postmodernist, or perhaps more accurately post-structuralist, theories.

6. Susanna Egan, like Frank, recognizes the importance—perhaps even the necessity—of crisis in contemporary autobiographical writing (1999, 4).

7. In his book *Genesis and Development of a Scientific Fact*, first published in German in 1935, Ludwik Fleck investigates the means by which an empirical fact comes to be thought. In order to demonstrate the emergence of an empirical fact, he offers as an example not a history of syphilis, but a "history of the concept of syphilis" (1979 [1935], 11). Thinking about illness as an *experience* as well as an *event* connects to Fleck's rendering of a "history of the concept" of a particular illness.

8. This will relate to my discussion below of the use of rest as a form of discipline in the TB sanatorium: that is, rest becomes an extreme objectification of the body almost to the point of death. Interestingly, in his book *Complications* (2002), Atul Gawande discusses the fact that the practice of autopsy has become rare in the twenty-first century, not because people refuse to have their loved ones cut open after death, but because doctors rarely ask to perform them. Gawande believes that "what discourages autopsies is medicine's twenty-first century, tall-in-the-saddle-confidence," and its faith in the unlikelihood that a mistake in diagnosis or treatment will be found (193–94). In my conclusion I discuss Gawande's critique of medicine's confidence in its knowledges and procedures, a confidence that is unwarranted according to many studies that Gawande reviews.

9. Hawkins takes the term from Oliver Sacks's *Awakenings*. He, in turn, takes the term from Freud, who called his study of Leonardo da Vinci a "pathography." Pathographies, like Freud's case histories, are biographies or autobiographies that combine science and art. Hawkins defines pathography as "a form of autobiography or biography that describes personal experience of illness, treatment, and sometimes death" (1993, 1). Clearly, Hawkins would call the objects of investigation in *Treatments* "pathographies," though I myself do not favor this term because of its etymological link to "pathology."

10. This will be a main focus of the second half of chapter 4 where I analyze doctors' narratives of AIDS.

11. In a review of *Madonna Swan* in the *Journal of American History*, St. Pierre's role is effusively celebrated and not at all problematized. In fact, a significant portion of the review is about what the reviewer takes to be his heroic role in bringing this story to readers. The reviewer describes at length St. Pierre's role, lauding him for shaping Swan's personal anecdotes "into a unified vignette that captured the essence of Madonna's spirit" (Price 1992, 1224). The review begins and ends with St. Pierre; in a sense, his story gives shape not only to *Madonna Swan* but to the reviews of it as well. After asserting that *Madonna Swan* is a testament to remarkable women so often ignored in Native American history and biography, the reviewer ends by noting: "This book is also a testament, however, to the foresight and sensitivity of Mark St. Pierre himself, who, unlike other collaborators and 'interviewers,' had the courage to allow Madonna's words to speak loudly and clearly for themselves" (1224). One really must ask, as Gayatri Spivak has, "Can the Subaltern Speak?" (1994). This is a question I will again ask in chapter 4 when I discuss the narratives of AIDS doctors about their patients.

12. See, for example, Bates 1992, Craddock 2000, Dormandy 2000, Feldberg 1995, Ott 1996, and Rothman 1994.

13. For example, Craddock analyzes writings in *Hi-Life*, a newsletter written and produced by patients at Arequipa sanatorium outside San Francisco (2000, 184).

14. For an interesting discussion of the history of Firland, see the Online Encyclopedia of Washington State History at www.historylink.org.

15. This is a term I use to highlight the importance of the field of health and illness in understanding class differences, and which I derive from and in relation to Bourdieu's other forms of capital: economic, social, cultural, and symbolic (Bourdieu 1984). Health capital refers to access to material objects, including water, food, and housing; institutions, including preventive and emergency health care, health insurance, and health education; and relationships with trusted friends or family members who may act on one's behalf as a health advocate or authority in health care emergencies. I briefly mention the importance of relationships as a form of health capital again in my conclusion.

16. For an application of Foucault's theory/method in the area of health, see Rose 1994 (58).

17. I see the important early work on medicalization by Zola (1983) and others as related to, but not quite the same as, Foucault's theories of power. Medicalization might be understood as one technique of discipline. Zola argues that, "the full exercise and expansion of medicine's influence has awaited the mid-twentieth century" in his essay "Medicine as an Institution of Social Control" originally published in 1972 (1983, 258). In the next chapter, I will look at political challenges to this exercise and expansion of medical expertise.

18. For a Foucauldian analysis of the "reproduction of gender" norms in the nineteenth and twentieth centuries through the practices of diet and exercise, see Bordo 1990 and 1993, and Bartky 1990.

19. I discuss the proto-ethics of failure in detail in my conclusion.

20. This idea of letting particular populations die, ideally away from the "general population," anticipates what will happen with AIDS in the United States in the

early days of the epidemic when it was found mostly in gay male and IV–drug using populations.

21. He also could have been describing AIDS in Africa in the late twentieth and early twenty-first centuries. While AIDS in Africa is spoken of as an "epidemic," it has been approached, particularly by Western governments and organizations, as something that has become endemic, and therefore not cost effective to treat. For a discussion of the apparent permanence of death from AIDS in Africa, despite effective treatments, see Farmer 1999 and 2003.

22. According to Feldberg, "By the early 1930s, Americans had become critically aware of the extent to which tuberculosis was decimating Native American communities. The calls for 'Indian assimilation,' first made at the turn of the century, had taken their toll as Native Americans, both off the reservations and on, adopted an impoverished and crude facsimile of the 'white' lifestyle, which compromised their health" (1995, 168–69).

23. Recall that MacDonald remains at the Pines for nine months, whereas Madonna Swan is at Sioux San for six years.

24. Although the patients at Sioux San were able to get some fresh air through windows and balconies, the image one gets of Sioux San is contrary to the popular image that MacDonald alludes to of sanatoriums as places that promoted access to pure air for patients. For more general discussions of sanatorium life, see Dubos and Dubos 1952, Bates 1992, and Rothman 1994.

25. As Abel and Reifel note in their study of health care on reservations in the 1930s, "Tuberculosis was, in the words of the Merriam Commission, 'without a doubt the most serious disease among Indians.' The tuberculosis death rate was seven times that of the rest of the population" (1999, 497).

26. This experience—of witnessing one's own death in the dying of others—will come up again in my discussion in chapter 5 of Paul Monette's writing on the experience and event of AIDS in the gay community in the United States.

2. Politicizing Patienthood

1. "Subjectivation" is Robert Hurley's translation of the French word *asujettissement*, which comes from the verb *asujettir*, meaning "to subjectify." I prefer the translation "subjectification" because it relates grammatically to the English word "objectification." I have, however, maintained Hurley's translation in quotations from the English editions.

2. The individual who would form himself as an ethical subject of his actions in the period Foucault is describing is, of course, always free and always male. "It was," Foucault writes, "an ethics for men: an ethics thought, written, and taught by men, and addressed to men—to free men, obviously. A male ethics, consequently, in which women figured only as objects or, at most, as partners that one had best train, educate, and watch over when one had them under one's power, but stay away from when they were under the power of someone else (father, husband, tutor). This is doubtless one of the most remarkable aspects of that moral reflection: it did not try to define a field of conduct and a domain of valid rules—subject to the necessary modulations—from

the two sexes in common; it was an elaboration of masculine conduct carried out from the viewpoint of men in order to give form to *their* behavior" (1985, 22–23).

3. The appellation for Singer is from Judith Butler, who edited *Erotic Welfare* (1993), a collection of Singer's essays published after Singer's death from kidney cancer in 1990.

4. While the rhetoric and politics of AIDS served as a source of inspiration, it also served as a source of irritation for many persons with breast cancer: Why, they wondered, was so much government money being invested to fight AIDS in the early 1990s, when research on breast cancer—numerically affecting more people than AIDS in the United States, though, in the early days of the AIDS epidemic at least, not quite so lethal—was so underfunded? Although a genealogy of breast cancer activism is beyond the scope of this project, there have been several recent histories of breast cancer and breast cancer activism, including Lerner 2001, Leopold 2000, and Stabiner 1997. The sometimes testy relationship over funding priorities, partially as a result of the *disease du jour* mentality of the media and the public, isn't to say that many persons with breast cancer don't feel, on a personal level at the very least, a connection with persons with AIDS (PWAs), as will become apparent in my discussion of Eve Sedgwick's essay "White Glasses." In the next chapter in which I compare breast cancer narratives in the British and American contexts, I take up Ruth Picardie's *Before I Say Goodbye* (1998), in which one of Picardie's most frequent email correspondents is Jamie, a friend who is HIV+. She tells Jamie, "It feels good to have a friend who is sick too. God knows, I wish you weren't, but there's a level of connection that even the most supportive friends can't achieve. And I don't want to talk about it with the sad, bald fucks you meet in hospital the whole time" (6).

5. The January 26, 1978, issue of the *New York Review of Books* contains the first part of three excerpts that would become *Illness as Metaphor*. It is interesting to note, in light of my use of Foucault's methods of historical analysis as well as his archaeological and genealogical descriptions of the modern episteme, that in this same issue Clifford Geertz reviews the English translation of *Discipline and Punish*.

6. The difference between being published by a small feminist press in San Francisco (*The Cancer Journals*), and being published by, first, the *New York Review of Books* and, then, Farrar, Straus and Giroux (in hardcover) and Vintage (in paperback) (*Illness as Metaphor*), and being published by a university press (Eve Sedgwick's *Tendencies*) is pertinent to my analysis of the different rhetorical strategies of the three texts. I will discuss the influence of the second wave women's movement on Lorde's work in more detail below, but, for now, I just want to point to the influence of small presses on the emergence and distribution of subjugated knowledges.

7. I am not arguing that the depoliticization of patienthood is a new phenomenon, but that it reasserts itself in a particular way in the 1990s. Although in 1972 Irving Zola wrote that, "the labels health and illness are remarkable 'depoliticizers' of an issue" (1983, 266), by 1980 this would not be the case thanks in large part to feminism.

8. I should note that Sontag doesn't really comment on the transformation of the representations of TB in the twentieth century, of which MacDonald's memoir and Swan's oral history are examples, and which I use to render an image of the modern patient as a docile and intelligible body and medicine as one institution that

regularizes populations. Once the tubercle bacillus was discovered in 1882, and once it became known in 1943 that TB could be treated successfully with antibiotics, TB became a less mysterious disease. It also became de-aestheticized, as it became associated with the poor and immigrant populations living in the tenement squalor of large cities. MacDonald herself refutes these negative associations because she wants to distance herself from certain "criminal" elements of society. Yet, as I argue in chapter 1, she doesn't necessarily want to eliminate the us/them structure, as much as to ensure that she, though a tuberculosis patient, is on the right (read: moral) side of the divide. The fact that Sontag doesn't address twentieth-century representations of TB, however, doesn't contradict her general argument about moral meanings attached to illness; rather, it confirms that such moral meanings are not ahistorical.

9. For example, in his book on what he calls "postmodern illness," David Morris uses the first two sentences of Sontag's image as the epigraph for his chapter entitled "The Country of the Ill" (1998, 21). In his book, Morris discusses Sontag's work at length and is not at all confused about her argument, but, nonetheless, he finds her metaphoric image as compelling as her anti-metaphoric idea.

10. Thanks to Cindy Patton for this insight.

11. And of course Sontag's much-eulogized death from cancer in December 2004 makes her own experience of cancer hardly a secret to my readers. Just as I was completing the final draft of *Treatments*, David Rieff, Sontag's son and a writer himself, published an account in the *New York Times Magazine* about his mother's last, desperate attempts to successfully treat her third and final bout with cancer. What is especially poignant—and terrifying—about Rieff's essay is his description of Sontag's horror at her impending death (Rieff 2005).

12. Sontag uses the first person only at the very beginning and the very end of *Illness as Metaphor*, and even in those instances does not describe her personal experience of illness. After her metaphoric description of the kingdom of the sick, she explains, "I want to describe, not what it is really like to emigrate to the kingdom of the ill and live there, but the punitive or sentimental fantasies concocted about the situation: not real geography, but stereotypes of national character" (1978, 3). Sontag says that her inquiry is dedicated to the elucidation of "lurid metaphors" and "a liberation from them" (3). Only in the book's last sentence does the first person return when Sontag notes, "The cancer metaphor will be made obsolete, I would predict, long before the problems it has reflected so persuasively will be resolved" (85).

13. Some, like D. A. Miller, whose critique of Sontag I will discuss in more detail below, claim Sontag actually bolsters rather than critiques these damaging metaphors.

14. By comparing Sontag with Lorde and Sedgwick in this way, I do not mean to de-intellectualize Lorde's or Sedgwick's work. It isn't that *The Cancer Journals* or "White Glasses" don't do things with ideas, but that they are not presented as solely, or even mainly, about what ideas can do.

15. It is important to note that Miller is most troubled by the fact that Sontag wants to "move beyond the specifically gay bearings of AIDS metaphors" (214). Therefore, he argues, contra Sontag's attempts to de-homosexualize AIDS, that AIDS is "the disease of gayness itself," which is precisely the sort of metaphorization that Sontag would find so troubling. Why does Miller press so desperately for AIDS to remain centrally a gay disease? By emphasizing the gayness of AIDS, he believes he

might also emphasize its potential political and personal effects. But, ironically, this itself is a rather urbane politics that circumscribes the potential affinities that might emerge out of the reality of AIDS, affinities beyond AIDS and its gayness. Lorde and Sedgwick, as we shall see, will articulate an ethics that emerges out of one's particular situatedness, but that also moves out into the world, and, in a sense, beyond oneself, but beyond oneself in a way that, paradoxically, emphasizes attachment rather than detachment. Nonetheless, Miller rightly, it seems to me, decries Sontag's attempts—as a writer—to remain "unsituated" and "impeccably detached" (216).

16. "Biomythography" is a term Lorde coined to differentiate *Zami* from conventional autobiography. Replacing the "auto" with "myth" highlights Lorde's attempt to write not a history of an individual, autonomous self, but a partially imagined, partially remembered "mythos" of a community of women that is at the center of Lorde's life ("bio") and writing ("graphy").

17. Worldwide breast cancer rates have increased 26 percent since 1980. In industrialized countries, it is one of the leading causes of cancer death, and, in the developing world, it is the second highest cause of cancer mortality. Breast cancer strikes one in nine American women and one in twelve British women. Japan has the lowest rate of breast cancer in the world, with one in twenty women affected. Nonetheless, the incidence in Japan is increasing as well; Japan's rate has doubled in the past thirty years (Ward 1997, 41). It should be noted, however, that in the United States lung cancer deaths outnumber breast cancer deaths, and that heart disease is more likely to kill women than all cancers combined. According to the National Center for Health Statistics (www.cdc.gov/nchs/), in 1997, 370,000 women died from heart disease, while 250,000 women died from cancer. Only in women under 55 was breast cancer more deadly than lung cancer. However, more women feared breast cancer than either heart disease or any other cancer. Phyllis A. Wingo et al. speculate that breast cancer is more feared than heart disease because of the "potential for disfiguring surgery and chemotherapy" (2000, 1004).

18. For more on the women's health movement in the 1970s, see Ruzek 1978, Dresser 1996, Davis 1999, Baxandall and Gordon 2000, and Morgen 2002.

19. For an illuminating look at the tactics of AIDS activism and its relationship to feminist health activism, see the ACT UP Oral History Project online (www.actuporalhistory.org). Sarah Schulman conducts the interviews and emphasizes in her questions the relationship between ACT UP and the tactics and strategies of other social movements, as well as what was original about ACT UP.

20. Arthur Frank calls Lorde's work prophetic as well. He reads *The Cancer Journals* as a manifesto, "carrying demands for social action." According to Frank, writers of manifestos, like Lorde, see "[s]ociety as suppressing a truth about suffering, and that truth must be told. These writers do not want to go back to a former state of health, which is often viewed as naive illusion. They want to use suffering to move others forward with them" (1995, 121).

21. In my most recent work, I have begun to trace what I call a genealogy of feminist AIDS activism. The tentative title for this project is *Doing Queer Love*, which looks for continuities and discontinuities between feminist health activism of the 1970s and AIDS activism of the 1980s. I seek the echoes of feminism in the AIDS

movement in order to begin to theorize something else that emerged with AIDS activism, which I call a way of *doing queer love*. Those three words are all crucial to my analysis. I want to offer a genealogy of forms of doing, not forms of being, and in particular, forms of doing love as opposed to, or different from, forms of being in love. In order to do this, I will consider AIDS activism in terms of, as Foucault would have it, a way of life as opposed to an identity. What does it mean to *do queer* rather than *be queer*, and how might this doing create new forms of not only queer sexuality, but also queer love? My discussion of Sedgwick below will suggest some ways of doing queer, and might be considered a strand of this genealogy.

22. Highlighting the theme of survival in Lorde's work and life, a film about her is aptly titled *A Litany for Survival*. The filmmakers understand that Lorde's litany is as much for those who haven't survived as for those who have.

23. For a cultural history of the breast, see Yalom 1997.

24. For insightful analyses of narrative as an alternative form of treatment, see Couser 1997 (77) and Stacey 1997.

25. In *Mirror Talk* (1999), Susanna Egan offers an interesting reading of *The Cancer Journals* in which she notes that Lorde disrupts "the possibility of any single linear narrative" and "demonstrates her multiple roles, conveying her sense of the inadequacy of any one form of writing. By means of such generic interactions . . . Lorde also foregrounds her concern that committing herself to paper is problematic for so multiple a self in such critical times. The unstable body destabilizes perception, recognition, and understanding so that the written texts present Lorde's fears that the 'light would change before the word is out, the ink was dry'" (218). I agree with Egan's assertion that Lorde's use of multiple forms for describing her experience conveys a sense of the "inadequacy of any one form of writing," and, as I will show in my discussion of Paul Monette's work in chapter 5, this practice of trying out different forms for telling the same story is not uncommon in illness narratives. I think, however, that this is not simply a generic question, but a problem of the inadequacy of language itself in the face of illness and dying, a point I will also take up in my conclusion.

26. The question of why black women, who are 12 percent less likely than white women to get breast cancer, are at the same time 50 percent more likely than whites to die from the disease, is a question Lorde would have been concerned with, not the question of how to achieve personal happiness and well being. A National Cancer Institute study published in 2000 shows that African Americans are 50 percent more likely than whites to get breast cancer before they turn thirty-five, when tumors are more aggressive, as well as "twice as likely to receive the diagnosis at an advanced stage, when the cancer has spread and is more difficult to cure" (Jetter 2000, D5). In fact, a "racial mortality gap" has been identified, which has everything to do with "being poor," according to Dr. Otis W. Brawley, an oncologist and epidemiologist who directs the National Cancer Institute's Office of Special Populations and Research (D5). And, Brawley also notes, as Lorde revealed in *Zami*, "being black may be a surrogate for being poor" (D5).

Recently, a number of works have pointed to the link between cancer and a toxic environment and have described instances of environmental justice activism against negligent corporations, local and state governments, and the federal government.

See, for example, Cole and Foster 2001, Spears 1998, Steingraber 1997, and Williams 1992.

27. *How to Do Things with Words* is a posthumous publication of a series of lectures that Austin gave at Harvard in the 1950s.

28. Aside from "White Glasses," her work that deals most directly with her own illness is *A Dialogue on Love* (1999), a memoir about the relationship between doctor and patient in psychoanalysis. Sedgwick begins analysis after experiencing depression as a result of her diagnosis of and treatment for breast cancer. Sedgwick has also written an afterword to Gary Fisher's illness narrative, *Gary in Your Pocket* (1996), in which she considers critically her role in editing and publishing Fisher's work after his death from AIDS. I will discuss Fisher's work and Sedgwick's role in getting it published briefly in chapter 4.

29. For an interesting discussion of the relationship between shame and desire, see Probyn 2005.

30. For an account of the connections between AIDS and breast cancer activism, see Boehmer 2000. Boehmer uses social movement theory and interviews with women AIDS and breast cancer activists to show the similarities and differences between the people involved in and the political strategies of these two movements. She maintains that the two movements started from "opposite vantage points" politically, because AIDS activists began as outsiders to the political arena and eventually became insiders, whereas breast cancer activists began as insiders, partly because of the success AIDS activists had in getting inside (21).

31. As Monica Pearl notes, Sedgwick "is rendered queer most notably through her identification with her friend Michael Lynch, an identification that takes place in the scene of sexuality but also, significantly, in the realm of illness. That is, Eve Sedgwick is queer because of her identification with Michael Lynch, the dying—but not yet dead—friend of her obituary" (2003, 63).

32. For a fascinating discussion of Michael Lynch's AIDS activism in particular and AIDS activism in Toronto more generally, see Silversides 2003.

33. Lorde discusses the importance of the two-stage operation in *The Cancer Journals*. Although before she had a biopsy she had already decided to have a mastectomy if her tumor proved malignant, she still wanted to proceed in two stages. She notes, "I wanted time to re-examine my decision, to search really for some other alternative that would give me good reasons to change my mind" (1980, 29). For a further discussion of the push for the two-step procedure, see Kushner 1975.

34. Some feminist health activist groups involved in contemporary breast cancer politics include the Women's Community Cancer Project in Cambridge, Massachusetts, the Women's Cancer Resource Center in Berkeley, and Breast Cancer Action in San Francisco.

35. My assertion that the politicization of patienthood has been covered over by a discourse of personal responsibility resonates with other recent accounts of the rise of neoliberalism, including essays in the special issue of *Public Culture* on "Millennial Capitalism and the Culture of Neoliberalism" (Comaroff 2000).

36. Including Zillah Eisenstein, whose book *Manmade Breast Cancers* (2001) makes many of the same arguments Ehrenreich makes. I should note that the immense

popularity of Ehrenreich's piece suggests politicization has not been entirely drowned out, though it is difficult to gauge whether the appeal of the piece is for its feminist politics or for its witty portrait of breast-cancer culture's infantilization of women.

3. Stories for and against the Self

1. For a detailed description of the "Bristol Approach," see Daniel 2000, and for a critique of alternative health practices in the British context, see Coward 1989.

2. In chapter 5, I will look further at the defamiliarization of language in and through the experiences and events of AIDS and Alzheimer's disease.

3. In my conclusion I will discuss medical infallibility within the idiom of malpractice. Following Atul Gawande's discussion of medical error, I suggest that error creates a *différend* in medicine that can only be addressed through malpractice and/or media scandal, which in fact does not really address the problem. Rosenblum's and Picardie's cases seem unequivocable and inexcusable, which makes them appear especially scandalous.

4. In his foreword, Matt Seaton notes that email was convenient for Picardie when she was ill, and "a way of expressing thoughts and feelings more spontaneously than in a letter, yet more reflectively than in a telephone conversation. It had a quality of being simultaneously intimate and serious, yet transient and disposable, and this meshed with something in her writer's psyche" (viii). As Seaton indicates, the email form captures well what I describe below as Picardie's "passionate irony": that is, her consistent opposition to sentiment combined with a desire to make connections despite her cancer's capacity to isolate her. Or, as Picardie herself states in an email to her friend India Knight, "I like cybercancer better than phone-cancer—makes me less prone to weeping, boredom, self-pity, etc." (13).

5. It is important to keep in mind that both *Cancer in Two Voices* and *Before I Say Goodbye*, unlike, for example, Lorde's *Cancer Journals* or Sedgwick's "White Glasses," end with the death of the person with breast cancer, as well as with each partner's attempts to practice and articulate mourning in the face of immense loss. Clearly, not all illness narratives end in death, and I will take up this point further in my discussion in the next chapter of *A Leg to Stand On*, by Oliver Sacks. Frank calls stories of illness and recovery "restitution stories." He notes that the restitution story is, in fact, the most common illness narrative because people who are sick want to be healthy again and because society "treats health as the normal condition that people ought to have restored" (1995, 77). Thomas Couser agrees with Frank, arguing that the plot of breast cancer narratives "is a comic one; it ends 'happily,' with some significant recovery; the narrators are healed, if not cured. Without exception, then, the narrators are, or claim to be, better off at the end than at the beginning" (1997, 39). Later, Couser does distinguish "breast cancer diaries" from "breast cancer narratives," and acknowledges that the former often do not end happily and may be published after the death of the author. As I hope to show, Couser is describing what he understands as a general narrative phenomenon, but which I take to be specific to an American model of illness narratives. Picardie, as we shall see, does not claim to be better off because of her cancer, or even because of writing her narrative of cancer.

6. For further analysis of bodily health as a manifestation of spiritual and moral health, see Zola 1983, Crawford 1984, and Coward 1989.

7. Including, for example, Ablom 1997; Williams 1992; Wittman 1993; Campo 1997; Verghese 1994; Monette 1988a and 1988b; Reeve 1988; and Doty 1996.

8. It is important to note, especially since Franklin's *Autobiography* is considered the quintessential American autobiography and is linked, by Benedict Anderson among others, with the formation of the American nation, that Franklin's *Autobiography* was originally published in French in 1791 with the title *Mémoires de la Vie Privée* and only translated into English and published in London two years later.

9. For a discussion of the place of autobiography in the American literary tradition, see Parini 1998. Parini notes that "[a]utobiography might well be called the essential American genre, a form of writing closely allied to our own national self-consciousness" (A40).

10. Nancy Mairs, who, as I mentioned in my introduction, was one of the first people to point to the phenomenon of what she calls "the literature of personal disaster" (1994), is clearly one exception to my American model. In her collection of autobiographical essays, *Waist-High in the World* (1996), about the experience of living with multiple sclerosis, she writes: "This is not . . . a 'feel-good' book. Too bad, because if it were, it might spend weeks on the *New York Times* best-seller list and make of me the wealthy woman I have always dreamed of being. In truth, I would give my eyeteeth (which narrowly escaped being knocked out along with my incisors when I fell flat on my face several years back) to write a best-seller, but this is not it. I am not now, nor have I ever been, a member of the inspirational class. I suppose you might call this instead a 'feel-real' book, and reality has never been high on any popular list. I ask you to read this book, then, not to be uplifted, but to be lowered and steadied into what may be unfamiliar, but is not inhospitable, space. Sink down beside me, take my hand, and together we'll watch the waists of the world drift past." (18)

11. See, for example, Bayley 1998, 2000a, and 2001; Diamond 1998; Stacey 1997; Rose 1995; Segrave 1995; and Watt 1996. I will discuss Bayley's writings on Iris Murdoch's Alzheimer's in chapter 5 and Rose's *Love's Work* in my conclusion.

12. These are Frayn's words. He tells Seaton, "Even in the things that look most frivolous there has to be the threat of something quite painful to make the comedy work. . . . I suppose the play of mine that's best known is 'Noises Off,' which everyone thinks is a simple farce about actors making fools of themselves. But I think it makes people laugh because everyone is terrified inside themselves of having some kind of breakdown, of being unable to go on. When people laugh at that play, they're laughing at a surrogate version of the disaster which might occur to them" (Seaton 1999, 8).

13. The main point of Sullivan's article, it should be noted, is to assert that recently "the islanders have . . . sensed that there is only so long, even if you are English, that you can cling to a culture of loss. Sooner or later, you begin to feel the possibility of gain" (1999, 79). One might wonder if Sullivan describes a shift in national *zeitgeist* or the shift in his own sensibility as he has transplanted himself from one side of the pond to the other. Or, it may be that what I describe here as a British mode of being ill is a manifestation of efforts to "cling to a culture of loss," because this contemporary culture of loss also signifies, implicitly at least, a supposedly heroic imperial past.

14. Which isn't to say that there haven't been illness narratives published in American newspapers. See, for example, Cathy Hainer's columns on breast cancer published in *USA Today* from 10 March 1998 to 6 December 1999. I also discuss below Kim Stacy's column for the Owensboro (Kentucky) *Messenger-Inquirer,* but this cannot be said to have national circulation, as do the British newspapers mentioned here and *USA Today*. And this, it seems to me, is the important point: that most of the major British newspapers have nationwide circulation, whereas most major American newspapers, except perhaps the *New York Times* and *USA Today,* do not.

15. For further reports on the Stacy story, see Kuczynski 1999 (18) and Roeper 1999 (11).

16. Elisa Segrave also comments on what she understands as an American model of political and personal agency. Because she spends a fair amount of time in the States with her "American cousins," she comes in contact with both HIV/AIDS and breast cancer activism in the United States. She notes enthusiastically, "I like the way the Americans tackle this kind of thing head-on. I like the way in this country there is still the feeling that one can influence one's fate, that the lay person can still do something. Also, people here seem more open about illness." But Segrave knows the downside of the situation in the States and, in a parenthetical remark, takes aim at what can be disastrous about being ill in America: "Of course, if you're not insured, you can be ruined. American doctors cost the earth" (1995, 168).

17. Although Picardie is willing to combine conventional and complementary medicine (as she calls alternative medicine), British writers in general are much more skeptical about alternative medicine (see Coward 1989; Rose 1995; Diamond 1998; and Stacey 1997). This is perhaps in part because of the association of alternative medicine with American self-help culture, but it also seems to have something to do with the contrast between the more passive approach of the British National Health Service (NHS) and the more active, interventionist approach of American medicine, which means the American approach often becomes, in Britain, the sort of alternative one seeks, rather than a holistic, noninterventionist alternative approach.

18. In an interesting reading of Picardie's *Observer* columns and Kathy Acker's article in the *Guardian* entitled "The Gift of Disease," published in 1997, Saywell, Beattie, and Henderson describe these narratives as "alternative media accounts of breast cancer" (2000, 54). They show how Picardie's and Acker's accounts diverge from mainstream media coverage of breast cancer in Britain that "construct[s] the images and experiences of breast cancer within the confines of acceptable definitions of femininity and illness" (54). Their analysis supports my suggestion that it is possible for a personal narrative of illness to provide a much-needed counternarrative to mainstream media and medical accounts.

19. I take this term from Bersani's discussion of Mallarmé and sublimation in *The Freudian Body* (1986). According to Bersani, "Mallarmé encourages us to view sublimation not as a mechanism by which desire is denied, but rather as a self-reflexive activity by which desire multiplies and diversifies its representations" (49). For Bersani, Mallarmé's irony is "additive rather than corrosive" (49), which I think could be said for Picardie's irony, if not for Geete's.

20. And nor does Lorde's later diary, *A Burst of Light* (1988), which takes up her situation after her breast cancer metastasizes to her liver in 1984.

21. Among the tributes only one sounds a note of disquiet about Lorde's legacy. Linda Cue, a black, feminist writer from Gainesville, Florida, writes that upon reading that Lorde is dead, she stays up "almost the entire night hoping there would maybe be something about her death on the news. Maybe CNN, I thought, watching the TV almost all night before falling asleep. There was nothing, or maybe they announced her death after I fell asleep" (88).

22. For an informative study on contemporary elegists who seek to offer their readers something "beyond consolation," see Zeiger 1997.

23. Missing the death of another comes up in several of the texts I take up in *Treatments*, including Rafael Campo's work discussed in chapter 4 and Paul Monette's work discussed in chapter 5. As I will highlight in my discussion of Monette, such missed encounters demonstrate what Lacan called the real.

24. I will return to Scarry's account of the experience of pain in my conclusion.

25. In *Mirror Talk* (1999) Susanna Egan compares Butler and Rosenblum's text and its "interactive procedure" to Ken Wilber's *Grace and Grit* (1991), which describes the death of his wife, Treya, and includes excerpts from her journals. According to Egan, "Her writing interacts with his, the two of them developing a line of thought or narrative as if in a dialogue. Vertical lines in the left margin identify Treya's journal voice and distinguish that private expression from her public talks and letters" (219). Egan understands that Wilber's role, like Butler's and Seaton's, as final narrator is "problematic," but she doesn't dwell on the very questionable appropriation of his wife's voice in what ostensibly becomes his book (he is listed as sole author of *Grace and Grit*). I return in further detail to the complex problem of appropriation in the next two chapters.

26. The words are from Elizabeth Spelman's *Fruits of Sorrow* (1997). In opposition to the problem of the appropriation of an-other's suffering, Spelman acknowledges that we can only make sense of suffering by paying a certain *attention* to it. Spelman's understanding of attention is taken from the work of French philosopher Simone Weil, who described attention as a form of obedience to the other, a prayerful sundering of the self in relation to the other. Weil's understanding of attention is similar in many respects to Emmanuel Levinas's understanding that the relationship to the other is an ethical one; that is, the other is a summons or call to an ethical relationship.

4. Becoming-Patient

1. For a discussion of Sacks in relation to other writers, including Nancy Mairs and Jean-Dominique Bauby, who I also call phenomenologists of disability, see Diedrich 2001. Sacks, it should be noted, reads philosophy; he spices his works with quotes from philosophers (in *A Leg to Stand On*, among the most frequently cited are Wittgenstein, Nietzsche, and Kant), and provides discussions of philosophical questions. The original version of *A Leg to Stand On*, published in 1984, ends with Sacks's arrival at something he calls "understanding," an understanding of the experience he has had that is not just or simply clinical, but also "existential." Sacks relates his understanding to Kant's notion of the "synthetic *a priori*," or what he calls more generally "a

transcendental, Kantian science" (1993, 182). His own interpretation of his experience of coming to understanding seems somewhat odd to me, in that he has provided, up until that point, a plainly phenomenological description of body-image breakdown that results from the injury to his left leg. In an afterword to an edition published ten years after the original publication, it is apparent that Sacks has become familiar with phenomenology—though he does not use the term—and he rejects his original excursion into what he himself now calls Kantian "mysticism" in order to discuss the relationship between neurology and consciousness. Sacks seems to call for a phenomenological neurology when he writes, "The structure of consciousness, in general, has not been approached by neurologists until very recently; they have felt, all too often, that consciousness is not their business, but a matter best left to psychiatrists. This, indeed, is a relic of the severe dualism of the last century, which divided phenomena into 'physical' or 'mental.' It was here, in this previously inadmissible space, that Babinski made his claim for a 'third realm'—a realm where organic, objective neurological disorders might nonetheless give rise to disorders of consciousness" (197).

Babinski's "third realm" sounds a lot like Merleau-Pontian phenomenology with its emphasis on consciousness as embodied. For an interesting reading of Sacks that also makes much of Sacks's shift from Kantian mysticism to the relationship between neurology and consciousness, specifically with reference to the cognitive science of, among others, neurologist Gerald M. Edelman, see Eakin 1999.

2. For a description of Head's experiments on himself, see Head 1920. For other discussions of Head's work, see Grosz 1994, especially the chapter, "Body Images: Neurophysiology and Corporeal Mappings," as well as the fictionalized account in Pat Barker's novel *Regeneration* (1991).

3. Sacks's story is a "restitution" story, in the narrative typology provided by Arthur Frank. Frank describes "restitution stories" as those narratives that project a "future that will not be disrupted by illness" (1995, 90). According to Frank, "The restitution story, precisely because it treats sickness as banal, displays a heroism in the face of bodily breakdown. But this heroism of the ill person is invariably tied to the more active heroism of the healer" (93). Sacks's own position as a patient who is also a doctor is interesting in this regard. He does not, in fact, treat his own disability as banal, and he is terrified at the strangeness of his own body and having to assume the passive role of patient, rather than the active, heroic role of healer.

4. Robert Murphy also describes this dissociation in cases, like his own, where paralysis is not localized to a single limb: "I have . . . become rather emotionally detached from my body, often referring to one of my limbs as *the* leg or *the* arm. People who help me on a regular basis have also fallen into this pattern ('I'll hold the arms and you grab the legs'), as if this depersonalization would compensate for what otherwise would be an intolerable violation of my personal space" (1987, 100).

S. Kay Toombs notes as well that the individual with multiple sclerosis often loses kinesthetic sensations, which, she explains, "not only give the body an 'interior,' clearly identifying it as 'mine,' but combine with movements in such a way that I experience such movements as my own" (1992, 132–33). In situations in which there is a loss of kinesthetic sensation, one is unable to perceive one's body as both a coherent whole and as one's own. See also Hockenberry (1995), who describes a serious burn to one of his paralyzed legs that he cannot feel. Hockenberry writes, "I had no sensation, yet I worried about the leg. I felt sorry for its pain. It could not tell me how it felt, it could not do anything for me" (45).

5. See also Carr 1986 (111); Hammond et al. 1991 (83); and Young 1990 (59).

6. In *The Body in the Mind*, Mark Johnson identifies "image schemata" as "recurring patterns without which our experience would be chaotic and incomprehensible." According to Johnson, these image schemata emerge from human bodily movement and positions, and then "can be figuratively developed and extended as . . . structure[s] around which meaning is organized at more abstract levels of cognition" (1987, ix–x). Thus, for example, "uprightness" comes to mean "upstanding" and "moral."

7. Many feminist philosophers have read Merleau-Ponty's lived-body-in-the-world as representing not a universal body, but a specifically male body (see, for example, Butler 1989; Grosz 1994; Shildrick 1997; and Young 1990). In her well-known essay "Throwing Like a Girl," Iris Marion Young suggests "that the modalities of feminine bodily comportment, mobility, and spatiality exhibit [a] tension between transcendence and immanence, between subjectivity and being a mere object" (1990, 55). Here Young is describing Merleau-Ponty's two bodies—the "phenomenal" and the "objective"—and placing the feminine-body-in-the-world as somewhere in between, or as occupying both positions at the same time. All the characteristics Young associates with feminine motility—"an *ambiguous transcendence*, an *inhibited intentionality*, and a *discontinuous unity* with its surroundings"—could just as easily describe the disabled-body-in-the-world (58, emphasis in original; see also Thomson 1997). The disabled person, in other words, is not only made passive, but is feminized as well. This is not to say that the experience of physical disability is *the same as* being female in the world, and I think Young is wrong to conflate the two as she does at the end of "Throwing Like a Girl" (65).

8. Interestingly, it is during the 1980s that Sacks is writing *A Leg to Stand On*, though he is describing a period of his life before the AIDS epidemic.

9. In a chapter in *The Renewal of Generosity*, Arthur W. Frank reads Campo's and Verghese's narratives—along with narratives by two other doctors, Lori Alvord (1999) and David Hilfiker (1985)—as examples of the possibility of physicians, who have been "demoralized by medical training and conditions of practice," being remoralized through storytelling (Frank 2004, 81). Frank suggests that *the practice of identification with the ill* becomes a key method in the remoralization of doctors. I think Frank's notion of the practice of the identification with the ill is related to my concept of becoming-patient. Although Frank recognizes that the practice of identification is an ongoing process that never reaches completion, he is not as concerned as I am about the possibility that the practice of identification with the ill might become a practice of appropriation.

10. Similarly, in Morris 1998 the author calls for a postmodern "biocultural model" to replace the modern biomedical model of medicine.

11. With the arrival of the protease inhibitors in 1995, HIV/AIDS has become more chronic in character, in the West at least where these drugs are readily available.

12. "Doctor/Outsider" is an adaptation of Audre Lorde's term "Sister/Outsider" (1984). Lorde uses the term to describe the ways, as a black lesbian, she both does and does not belong to the "black community" and the "lesbian community." Because of the hegemonic politics of sexuality in the "black community," she does not fit in as a

lesbian, and because of the hegemonic politics of race in the "lesbian community," she does not fit in as a black woman. I place "black community" and "lesbian community" in quotations because, of course, Lorde seeks to challenge such hegemonic articulations of belonging, and also understands these communities as imagined.

13. The fact that Verghese begins with his own imaginative re-creation of a patient's return to Johnson City and his final days there will be relevant in the last part of this chapter when I consider the question of appropriation in doctors' narratives. It is important to keep in mind at this point that Verghese, not the person with AIDS returning home, is telling this story. Thanks to Cindy Patton for pointing this out.

14. John Howard's important oral history of queer Mississippi, *Men Like That* (1999), suggests that Verghese's paradigm is only part of the story. Howard shows queer Mississippians do not always have to leave to become queer, but create their own Southern queer forms of belonging.

15. See Deleuze and Guattari 1986 and 1987 and Braidotti 1994 for further theoretical discussions of "nomadism" and "nomadic subjectivity."

16. Or, at least, this is how Verghese himself perceives his position in Johnson City. And, of course, because Verghese is telling the story, we are not presented with evidence to the contrary.

17. In *Recovering Bodies*, Thomas Couser discusses the role of technology in the doctor-patient relationship. Citing Reiser 1978, Couser describes a shift from the seventeenth century, where a "patient's testimony" was the "main basis of diagnosis," to the present, where the patient's testimony is discounted in favor of the use of diagnostic tools that ascertain "increasingly detailed and 'objective' evidence of the body's internal workings." As Couser explains, "[t]he effect has been to extend the range and domain of the medical gaze, and to efface the skin as an obstacle to vision and thus to diagnosis" (1997, 21).

18. "Rhizomatic model" and "arborescent model" are terms from Deleuze and Guattari 1987. According to Deleuze and Guattari, an arborescent model is hierarchical and conservative; its logic is defined by the practices of tracing and reproduction. The rhizomatic model, on the other hand, doesn't trace or reproduce; rather, it maps and constructs the unconscious. In the introduction to *A Thousand Plateaus*, Deleuze and Guattari write, "For both statements and desires, the issue is never to reduce the unconscious or to interpret it or to make it signify according to a tree model. The issue is to *produce the unconscious*, and with it new statements, different desires: the rhizome is precisely this production of the unconscious" (18). In the case of treating and writing AIDS there is nothing to trace or reproduce; what Campo and Verghese must do is unlearn an old medical logic by mapping a new logic of different statements and desires that emerge out of treating and writing AIDS.

19. In his "alternative therapeutic approach," Kleinman also conceives of the doctor-patient relationship as a negotiation. He asserts that the "real challenge is for the physician to engage in negotiation with the patient as colleagues involved in care as collaboration" (1988, 242).

20. For a discussion of this care network and the emergence of expert knowledge within this network, see Patton 1989 and 1990 and Epstein 1996. For an analysis of the emergence of Gay Men's Health Crisis (GMHC) as the model for AIDS service organizations, see Kayal 1993 and Lopez and Getzel 1987.

21. Part of Vickie's transformation is the real and symbolic regrowth of her hair. When she first comes to Verghese, she is virtually bald due to a persistent anxiety that causes her to pluck her hair out strand by strand. By the end of the book, her hair has grown back, and she no longer hides it under a scarf.

22. For an excellent oral history of AIDS doctors, see Bayer and Oppenheimer 2000. Bayer and Oppenheimer explain that when doctors confronted AIDS, they also confronted "their own therapeutic impotence . . . their own fears of an unknown infectious agent," and the unprecedented, large-scale death of individuals in the prime of their lives (5). Bayer and Oppenheimer describe their book as a "collective memoir" of doctors who became witnesses to history (7).

23. Campo and Verghese are both pricked by needles that they have just used on PWAs, exemplifying the quite real risk to doctors and other health professionals treating HIV/AIDS. Nonetheless, HIV/AIDS is hardly the first or only illness to incur risks on health professionals, and both doctors discuss the hysteria—even, or especially, within the medical profession—about the potential transmission of HIV from patient to health professional (or vice versa) particularly in the early days of the epidemic when there was uncertainty concerning means of transmission.

24. Verghese's *infection* dream, consciously or unconsciously, recalls Freud's *injection* dream that inaugurates *The Interpretation of Dreams* (1965 [1900], 138–53), which is also a doctor's autobiographical narrative. Infection, or becoming-infected, might also be seen as parallel to counter-transference in psychoanalytic practice.

25. It is interesting that Verghese explains his insomnia in terms of a desire to ward off his becoming infected in his dreams, whereas, as I show in the next chapter, Paul Monette explains his own insomnia as his means of countering the realities of infectedness, his partner's and his own.

26. For more about the link between sexuality and AIDS, see my discussion of D. A. Miller's critique of Susan Sontag's *AIDS and Its Metaphors* in chapter 2.

27. See Thomson 1996 for a historical, literary, and cultural study of "freakery" in America.

28. See my discussion in chapter 1 on Foucault's "archeology of medical perception," *The Birth of the Clinic* (1973).

29. In Deleuzoguattarian terminology, a "body without organs" is a set of practices that constantly dismantles the self rather than finds it. In *A Thousand Plateaus*, Deleuze and Guattari write that BwO (in their shorthand) is "what remains when you take everything away. What you take away is precisely the phantasy, and signifiances and subjectification as a whole" (1987, 151). In Verghese's examination of Luther, what passes between them is not meaningful in the sense of performing a rational medical function, but as an affective intensity that makes Luther smile, mesmerizes his parents, and gives Verghese pleasure. Although Deleuze and Guattari state that one can never reach the BwO once and for all, they also recognize suffering as a means by which one is forever attaining, and yet never obtaining, it. They list five forms—or, more to the point, practices—of embodiment that provide the experiences and experimentations necessary to dismantle the self into a BwO: the hypochondriac body, the paranoid body, the schizo body, the drugged body, and the masochist body (150). To this list, I would add the ill body and, in particular, the terminally ill body.

30. See Murphy and Poirier 1993 for an excellent collection of essays on just this question.

31. I will return to this question in more detail when I discuss Paul Monette's work in chapter 5.

32. In their work, Deleuze and Guattari distinguish between "desiring productions," as those "assemblages" that create becomings and multiplicities, and "social productions," as those forces that can only trace or reproduce what is already established or known. With regards to medicine in particular, what is reproduced is the binary logic of the doctor-patient relationship.

33. In March 1999, however, Lucky Mazibuko, an HIV+ South African, began a weekly part-autobiographical, part-educational column for the *Sowetan*, the largest daily newspaper in South Africa, on living with HIV (Swarns 1999).

34. This is represented textually by Kincaid's habit of reporting her brother saying something and then in parentheses offering the "original": a transcription of Antiguan pidgin English. Kincaid suggests her own role is to serve as cultural translator.

35. Zuger's comment that "the impact of HIV on the gay community is well described" (x) is problematic. Although I agree with Zuger that gay males have dominated the portrayals of the experience of AIDS in the West, I also think her broad brush ignores the varieties of gay experiences of AIDS (in other words, there is not just one gay male experience of AIDS), as well as the many media representations of AIDS that have tried to appeal to a supposedly broader, more general audience by portraying the allegedly "innocent victims of AIDS," including, for example, Kimberly Bergalis (see Rom 1997), Ryan White (1991), and Arthur Ashe (1993). For analyses of media representations of AIDS and PWAs, see for example, Crimp 1988, Carter and Watney 1989, and Treichler 1999.

36. Perhaps, with the enormous popularity of reality TV, this "documentary" look at patients' lives is a possibility for the future.

37. Verghese's second memoir, *The Tennis Partner* (1998), further reveals his identification with danger, and again includes an undercurrent of homoeroticism. He befriends an Australian medical student, David Smith, who is a former tennis pro, and their friendship blossoms in competitive sessions on the tennis court. When Smith confides to Verghese that he has a history of drug abuse, Verghese is curious to know the details of a life that revolves around getting high. He parallels his drive with Smith's, though again, his exploration of drug abuse remains a dangerous (intellectual and literary) mission from which he will "miraculously emerge unscathed," though Smith will not.

38. This association with Sedgwick will be particularly significant for Fisher, whose work Sedgwick edits and has published after his death. I briefly discuss their relationship below. More generally, both Campo and Fisher acknowledge Sedgwick's inspiration in their writing, and this inspiration seems to have as much to do with narrative styles of imagining as with imagining a political community that might emerge out of various queer identifications and practices as discussed in chapter 2.

39. I discuss Lacan's ethics seminar and his concept of being between two deaths in more detail in the next chapter.

40. My readings of Campo and Verghese are influenced by Elisabeth Bronfen's work in *Over Her Dead Body: Death, Femininity, and the Aesthetic* (1992) on the

inevitable violence of representations of dying and death, because they rely on: "the safe position of a spectator ('voyeur') and . . . a fragmentation and idolisation of the body—i.e. a severing of the body from its real materiality and its historical context ('fetishism')" (44).

41. Interestingly, although *Gary in Your Pocket* came out the year before *The Poetry of Healing*, Campo does not mention it.

5. Between Two Deaths

1. Bayley's trilogy includes the titles *Iris: A Memoir of Iris Murdoch* (1998) (published in the United States as *Elegy for Iris*), *Iris and the Friends* (2000a) (published in the United States as *Iris and Her Friends*), and *Widower's House* (2001).

2. Not to mention a text like Zuger 1995, which claims to be a documentary of her patients' lives.

3. I discuss the multiple practices of witnessing to historical trauma in the work of W. G. Sebald in Diedrich 2007.

4. Monette works with fable in the posthumously published *Sanctuary: A Tale of Life in the Woods* (1997).

5. For a discussion of Monette's *Becoming a Man*, along with Audre Lorde's *Cancer Journals*, in relation to abjection, as theorized by Judith Butler and Julia Kristeva, see Kimmich 1998.

6. I have also used Oliver's work to read Terry Tempest Williams's ethics of witnessing. In her memoir *Refuge* (1992), Williams is a witness to two events: the death of her mother to breast cancer and the rise of the Great Salt Lake and its swamping of a bird refuge where Williams, a naturalist, works. Williams eventually learns the reason for the cancer that has affected almost all of the women in her family: this epidemic of cancer is a deadly legacy of the nuclear tests in the allegedly empty deserts of the southwestern United States in the 1950s stored in the bodies of the women in her family who were downwind of the explosions. See Diedrich 2003 as well as the other essays in Chandler and Goldthwaite 2003.

7. This definition of the traumatic event is not unlike Foucault's definition of one's archive. Although I don't have the space to do so here, it might be interesting to think about Foucault's concept of the "blank space" in relation to trauma and the Lacanian real. Butler says the question of a "suppressed psychoanalysis in Foucault" might best be explored through his notion of resistance (1997b, 87), and, as I argued in the introduction, I understand subjugated knowledges as forms of repression at a historical not individual level.

8. Joseph Cady notes that in his collection of poems, *Love Alone*, "Monette matches his harrowing content with a harrowing style by upsetting every conventional expectation of order an audience might bring to the text" (1993, 249). In the poem "Three Rings," Monette captures an image of trauma: "why the world though stopped like a car wreck keeps doubling back" (1988b, 33), and Cady understands that "Monette incarnates this total 'wrecking' of his world in a thoroughly 'wrecked' form, designed to subject his readers to an immersive 'wrecking' in turn" (1993, 250).

9. As I showed in chapter 2, Audre Lorde also understands survival in these terms.

10. Unlike Rafael Campo, in his essay "Fifteen Minutes After Gary Died," which I discussed in the last chapter, Monette doesn't feel guilty for missing Roger's death. In fact, there seems to be a will to miss it on Monette's part, but this is less a form of denial than an acknowledgement that Roger's death is also always an echo of his own death in the future and the fact of his survival in the present.

11. Lacking punctuation or stanza breaks, Monette's poems are difficult to read. Nonetheless, like most poetry, Monette's is most accessible and powerful when read aloud, a process that inevitably leaves the reader out of breath and gasping for air, which was Monette's intention as he explains in the preface to *Love Alone:* "I have let them stand as raw as they came. But because several friends have wished for a few commas or a stanza break here and there, I feel I should make a comment on their form. I don't mean them to be impregnable, though I admit I want to allow no escape, like a hospital room, or indeed a mortal illness" (1988b, xii).

12. The trope of exile into a parallel universe recurs in Monette's work. In an attempt to describe this sense of "radical separateness" and liminality that the experience of AIDS affords, Monette refers, throughout *Borrowed Time*, to AIDS as an exile "on the moon" (1988a). For a discussion of the trope of exile in illness narratives, see Hawkins 1993 (81), and in Holocaust testimony, see Langer 1991 (53).

13. The trope of the "half life" returns again in *Becoming a Man* (1992), which is subtitled "Half a Life Story" and portrays the trauma of the closet as another death in life, a social death before a biological one.

14. The comparison of the event of AIDS with the Holocaust is rather common in AIDS writing. Larry Kramer was one of the first to make the comparison in both *The Normal Heart* (1985), a play, and in *Reports from the Holocaust* (1994), a 1989 journalistic indictment against the United States government's inaction in the face of the AIDS crisis and a memoir of his own journey into activism. In *The Normal Heart*, Kramer, never one for subtlety, reminds his reader that American Jewish organizations quietly, and behind the scenes, tried to convince the American government to act on behalf of European Jewry during World War II, rather than loudly and aggressively protesting the government's unwillingness to intervene. He compares such attempts at quiet diplomacy with early AIDS activism (or, according to Kramer, the lack of activism), and argues that this passive strategy likewise did not prevent genocide (1985, 6). My own use of the important critical works of Lawrence Langer, Dori Laub, and others on Holocaust testimony is not meant to deny the historical and experiential specificity of the Jewish Holocaust. But I do wish to locate the echoes—both affective and effective—between these unique and historically situated traumas.

15. To reduce that inscription to a number as Monette does in the title of this essay also points to what he elsewhere describes as the "subtle enslavement of the numbers" (1988b, 22). I read Monette's essay as both an epitaph itself and an essay on epitaphs in the tradition of Wordsworth. Calling Wordsworth's *Essays Upon Epitaphs* an "exemplary autobiographical text," Paul de Man notes, "It requires no lengthy argument to stress the autobiographical components in a text which turns compulsively from an essay *upon* epitaphs to being itself an epitaph and, more specifically, the author's own monumental inscription or autobiography" (1984, 72).

16. In *The Writing of the Disaster*, Maurice Blanchot writes that, "the cry tends to exceed all language, even if it lends itself to recuperation as language effect. It is both

sudden and patient; it has the suddenness of the interminable torment which is always over already—the patience of the cry: it does not simply come to a halt, reduced to nonsense, yet it does remain outside of sense—a meaning infinitely suspended, decried, decipherable-indecipherable" (1995, 51). Blanchot's description of the cry echoes Murdoch's contention that *King Lear* would be far more unbearable if Lear's cries continued interminably at the end of the play, a point I will take up in my discussion of Bayley's writings about Murdoch and of Murdoch's moral philosophy.

17. But, even the recent treatments that have transformed HIV from a disease that inevitably leads to death to a disease that is "managed" have not solved the puzzle of its timing. Much of the enthusiasm over the new treatment options is tempered by a sense of uncertainty surrounding the time between the disappearance of HIV from the blood and its feared reappearance.

18. In a dialogue with Cathy Caruth and Thomas Keenan, AIDS activist and filmmaker Gregg Bordowitz remarks: "And when I see somebody getting sick, I don't say, 'That *could* be me,' I say, 'That *will* be me.' So it's very painful and sometimes intolerable" (Caruth and Keenan 1995, 267; emphasis in original).

19. Even my choice of the phrase "person with Alzheimer's" to describe people who have a particular disease is not an obvious or ahistorical category of identity. I follow the *political* choice made by people with AIDS (PWAs) to emphasize that they are more than their disease, and more than "sufferers," "victims," and even "patients."

20. In the second volume of his trilogy, Bayley himself uses the term "uncanny" to describe "Iris's resemblance to a three-year-old," and his expectation that she will "shrink to an appropriate size" (2000a, 184). I discuss the uncanny in relation to the experience of illness in more detail in chapter 3. As I noted in that chapter one of Freud's examples of something that is uncanny is a severed head. One of Murdoch's best-known novels is called *The Severed Head* (1976), and one of its main characters is a psychoanalyst.

21. Others not as close to Murdoch as Bayley have also gotten plenty of column inches out of Murdoch's death. Many people, it seems, want in on the "let me tell you about Iris" bandwagon. See, for example, poet Wendy Bardsley's article in the *Independent* on her sixteen-year correspondence with Murdoch (Bardsley 2000). The Cult of Iris grew even larger with the Hollywood movie based on Bayley's first two books, starring Dame Judi Dench as Murdoch.

22. The fact that there is a genetic component in many—if not all—cases of Alzheimer's means that it is possible for a person to care for a parent living with and dying of Alzheimer's and to know, or at least suspect, that one is seeing one's future dying. In *Hard to Forget* (2000), a story that combines histories of science in general and Alzheimer's disease in particular with the history of one family, Charles P. Pierce describes the "genome cowboys" who are searching for the genes that cause Alzheimer's as well as what this search means to someone whose father, three uncles, and one aunt have all died of Alzheimer's. Others, including David Shenk, discuss the fact that with improved methods of diagnosis some people can now know they have Alzheimer's before they are profoundly affected by it. Shenk calls these people "the new class of Alzheimer's sufferers, diagnosed so early as to still be able to speak for themselves, to eloquently describe their experience, and to champion their rights" (2001, 94–95).

23. In her novel *The Sea, the Sea* (1978), Murdoch has her narrator, Charles Arrowby, espouse many of the same positions she articulates in this philosophical essay.

24. For more on spurious shortcuts to instant wisdom, recall my reading of the Oprahfication of literature in chapter 3.

25. In her reading of Lacan's reading of *Antigone*, Judith Butler argues that, for Lacan, while *Antigone* "does involve purgation—or expiation— . . . it is not one that leads to the restoration of calm but rather to the continuation of irresolution" (2000, 48). According to Butler, Lacan's image of Antigone is an "image of irresolution," and this, for Lacan, is the position of Being itself (49). The image of Antigone as an image of irresolution is linked finally in Lacan to "tragic action" (49), which is remarkably similar to Murdoch's reading of *King Lear* as exemplary tragedy, which I discuss below.

26. This relates to what Slavoj Zizek, in his work on Lacan and popular culture, describes as Lacan's "metaphor of anamorphosis." According to Zizek, this phenomenon occurs when a detail of a picture viewed in a straightforward manner is blurred and unintelligible; but, when the same picture is looked at awry or at an angle, it becomes a clear and distinguishable shape (1991, 11). It is precisely by looking awry that we might catch a glimpse of the real.

27. This seems to be a particularly difficult project where Alzheimer's is concerned. In *Hard to Forget*, Charles Pierce describes a community service in Wilmington, North Carolina, which provides organized activities for people with Alzheimer's for a few hours every Friday. Olga Jones, a former Army nurse who organized the Friday Group, as it is called, explains to Pierce that the group is not only for the patients, but also for their caregivers: "For a few hours, they're just not . . . *on duty*, you know?" (2000, 154) For Jones, seeing the ravages of Alzheimer's is worse than seeing cancer patients in the hospital or people severely wounded in battle: "These people disappeared while you looked at them" (155). In acknowledgement of these disappeared selves, Jones calls the people she works with "Alzheimers"; they have, in essence, according to Jones's form of naming, become their disease. Although this is not quite the same thing as saying that they are a corpse, as Bayley's acquaintance does, it is also quite different from an attempt to de-link them from their disease through naming, as the distinction between AIDS victim or sufferer and person with AIDS exemplifies. Although Bayley, to a great extent, traces Murdoch's becoming Alzheimers, he nonetheless always also seeks to assert her individuality.

28. The term "proximate other" is from Eakin 1999.

29. The phrase "limits to representation and representability" is from Butler 2000.

Conclusion

1. I briefly referred to Scarry's work in chapter 3 in relation to Barbara Rosenblum's mother's wailing at her daughter's funeral.

2. For a discussion of AIDS and other emergent infectious diseases, see Garrett 1994. Garrett describes the 1950s and 1960s as a period of "tremendous optimism" in the battle against infectious diseases, partly because of the discovery and

improvement of antibiotic therapies, and partly because of a faith in progress and what she calls "health transition": the notion that "as nations moved out of poverty and the basic food and housing needs of the populations were met, scientists could use the pharmaceutical and chemical tools at hand to wipe out parasites, bacteria, and viruses" (31).

3. I'd like to thank the anonymous reader of *Treatments* who helped me see this important point regarding the relationship between failure and medicine.

4. I take the term "empirical philosophy" from Annemarie Mol. In her book *The Body Multiple* (2002), Mol is concerned not with how medicine knows a particular disease, or how a patient knows her illness, but with how that disease is enacted through practices. Mol calls her work an "empirical philosophy" of atherosclerosis that "foregrounds practicalities, materialities, *events*" (12–13; emphasis in original). Illness, for Mol, "is something being done to you, the patient. And something that, as a patient, you do" (20). This relates to my concept of arts of being ill and doing illness.

5. The Institute of Medicine's 1999 report on error in medicine bears this out. That report, published as *To Err Is Human* in 2000, found that between 44,000 and 98,000 people die each year in American hospitals from preventable medical errors.

6. By saying this, I do not mean to imply that doctors who make themselves available to treat the family members of colleagues are acting unethically. In fact, this is often done on their own time, and sometimes at their own expense, and seems to represent an ethical practice in medicine that shouldn't be eliminated but taken account of when we consider inequalities in access to good care. I myself have benefited from this practice, when I was fit in for emergency neurosurgery to remove a ruptured disk in my neck that was compressing my spinal cord because my sister is a neurologist and took care of everything for me. The day after my sister examined me, and knew something was very wrong, I was at the hospital where she works at 7 a.m. having an MRI. When I thanked the tech for fitting me in, she admitted that they were only doing it because I was Dr. Diedrich's sister, which I think says that the way my sister practices medicine is regarded, by her colleagues at least, as ethical.

7. Thanks to Victoria Hesford for clarifying this point for me.

8. Gawande's suggestion that the notion that error is intolerable is an American idea relates to my discussion in chapter 3 of national arts of being ill.

9. In their introduction to psychologist Silvan Tomkins's work on affect, Eve Kosofsky Sedgwick and Adam Frank discuss the way Tomkins turned to systems theory to provide a theoretical model for the affects. They explain that, "[t]he valorization of feedback in systems theory is . . . necessarily a valorization of error and blindness as productive of, specifically, *structure*" (1995, 13; emphasis in original). This is precisely the point I am trying to make about what an ethics of failure might be and do, as articulated in Gawande's work. This also relates to Georges Canguilhem's theory of error. In his introduction to a reissue of Canguilhem's essays on the normal and the pathological, Foucault discusses Canguilhem's borrowings from early information theory to articulate a theory of error in the life sciences. Foucault writes, "At the heart of these problems [of evolution] is that of error. For at life's most basic level, the play of code and decoding leaves room for chance, which, before being disease, deficit or monstrosity, is something like perturbation in the information system, something

like a 'mistake.' In the extreme, life is what is capable of error" (Canguilhem 1991, 21–22). I thank the anonymous reader of *Treatments* who pushed me to think about Canguilhem in relation to my concept of an ethics of failure. As that reader noted, Canguilhem argues that, "To be sick means that a man really lives another life, even in the biological sense of the word" (88).

10. Here I am suggesting an interdisciplinary science studies approach to the practice of medicine, in the tradition of Latour and Woolgar 1986, Latour 1993, Law 1994, Mol 2002, and Law and Mol 2002. Again, Canguilhem's early analysis of the multiple practices of medicine is pertinent here. Like Gawande, Canguilhem before him saw "scientific rationalization" as an important aspect of, but not at all the whole of, what medicine is, can, and should be. According to Canguilhem, "despite so many laudable efforts to introduce methods of scientific rationalization, the essential lay in the clinic and therapeutics, that is, in a technique of establishing and restoring the normal which cannot be reduced entirely and simply to a single form of knowledge" (1991, 34).

11. Although I don't discount this explanation, I was somewhat surprised when I taught a film class to a small group of medical students as part of the medical humanities program at Stony Brook that some of the students spoke quite openly about certain specialties as being desirable because they required minimal contact with patients and their suffering.

12. Recall my discussion in chapter 3 of Ruth Picardie, who also experienced her pregnant body as averse, not from nausea so much as from the changes in her body that came with pregnancy and the feeling of having something alien to the self within the self, which Picardie says is similar to the relationship she has to her cancer.

13. Just as I was completing this chapter, an essay appeared in the *New York Times* by the doctor Richard Friedman entitled "Learning Words They Rarely Teach in Medical School: 'I'm Sorry'" (2005). In the essay, Friedman describes a patient who complained to him about the treatment she had received from a resident Friedman supervises. The psychiatric resident had not told her that a new medication he prescribed for her would have adverse side effects if taken with other medications she was currently taking. Friedman explains that, "What really bothered her was not so much the physical discomfort—headache and dizziness—but the fact that when she confronted her doctor . . . he didn't apologize for his mistake." When Friedman asks the resident why he didn't apologize, he is surprised to discover it is not because he felt his "medical authority was threatened" or "the legal implications of admitting a mistake," but that "acknowledging the mistake would narrow the psychological distance between him and his patient, and that felt uncomfortable" (5).

14. Thanks to Cindy Patton and the Global Science/Women's Health working group for helping me develop this idea.

15. For further articulation of a feminist ethic of care, see for example, Noddings 1984, Tronto 1993, Wendell 1996, Kittay and Feder 2003, and Hankivsky 2004. For feminist critiques of the ethic of care, see for example, Diprose 1994 and Shildrick 1997.

16. Cities in Rose's work stand in for aesthetic, philosophical, and ethico-political ideas. In *Mourning Becomes the Law* (1996), Athens represents rational politics, what Rose calls the "New" Jerusalem represents the imaginary community of a new ethics, and Auschwitz represents the irrational.

17. The question of whether AIDS is overspoken is certainly debatable. As I noted in chapter 4, the question of whose experience of AIDS is spoken and whose is not is important to keep in mind. Having said that, I would agree with Rose that the experience of ovarian cancer, especially in 1995 if less so now, is not overspoken. For other autobiographical accounts of ovarian cancer, see Radner 1989 and Tilberis 1998. And, of course, in the immensely popular play and HBO film *Wit*, the protagonist, Vivian Bearing, is treated for and dies of ovarian cancer (Edson 1999).

18. Rose, whose Jewish father and mother emigrated to England from Poland before World War II, and who lost numerous family members in the Holocaust, interweaves the stories of her three "Cities of Death" (34)—New York, Auschwitz, and Jerusalem—throughout *Love's Work*.

19. The quotation is from Staretz Silouan (1866–1938).

20. Although Rose portrays them as similar, the two also appear, in Rose's view, to be mutually exclusive philosophical and political positions.

21. This was published after Rose's death, and contains essays that are more straightforwardly philosophical and less literary and personal than *Love's Work*. Nonetheless, these essays are particularly revealing read in light of *Love's Work*.

Bibliography

Abel, Emily K., and Nancy Reifel. 1999. "Interactions between Public Health Nurses and Clients on American Indian Reservations During the 1930s." In *Women and Health in America*. 2nd ed., ed. Judith Walzer Leavitt, 489–506. Madison: University of Wisconsin Press.

Acker, Kathy. 1997. "The Gift of Disease." *Guardian*, 18 January.

Adams, Henry. 1974 [1906]. *The Education of Henry Adams*. Boston: Houghton Mifflin.

Albom, Mitch. 1997. *Tuesdays with Morrie: An Old Man, a Young Man, and Life's Greatest Lesson*. New York: Doubleday.

Alvord, Lori Arviso, with Elizabeth Cohen Van Pelt. 1999. *The Scalpel and the Silver Bear: The First Navajo Woman Surgeon Combines Western Medicine and Traditional Healing*. New York: Norton.

Amis, Martin. 2000. *Experience*. New York: Hyperion.

Anderson, Benedict. 1991. *Imagined Communities: Reflections on the Origin and Spread of Nationalism*. Rev. ed. London and New York: Verso.

Ansen. 1993. "AIDS and the Arts: A Lost Generation." *Newsweek*, 18 January, 16–23.

Armstrong, Stephen. 1999. "Dead Funny." *Sunday Times* (London), 27 June, 10.

Ashe, Arthur, and Arnold Rampersand. 1993. *Days of Grace: A Memoir*. New York: Alfred A. Knopf.

Atlas, James. 1996. "The Age of the Literary Memoir Is Now." *New York Times Magazine*, 12 May, 25–27.

Austin, J. L. 1961. *How to Do Things with Words*. Cambridge, Mass.: Harvard University Press.

Bardsley, Wendy. 2000. "Dear Iris . . . With Much Love, Wendy." *Independent* (London), 24 March, 7.

Barker, Pat. 1991. *Regeneration*. New York: Dutton.

Bartky, Sandra Lee. 1990. *Femininity and Domination: Studies in the Phenomenology of Oppression*. New York: Routledge.

Bates, Barbara. 1992. *Bargaining for Life: A Social History of Tuberculosis, 1876–1938*. Philadelphia: University of Pennsylvania Press.

Batt, Sharon. 1994. *Patient No More: The Politics of Breast Cancer*. Charlottetown, Prince Edward Island: Gynergy Books.

———. 1998. "'Perfect People': Cancer Charities." In *The Politics of Women's Bodies: Sexuality, Appearance, and Behavior*, ed. Rose Weitz, 137–46. New York and Oxford: Oxford University Press.

Bauby, Jean-Dominique. 1997. *The Diving Bell and the Butterfly*. Trans. Jeremy Leggatt. New York: Alfred A. Knopf.

Baxandall, Rosalyn, and Linda Gordon. 2000. *Dear Sisters: Dispatches from the Women's Liberation Movement*. New York: Basic Books.

Bayer, Ronald, and Gerald M. Oppenheimer. 2000. *AIDS Doctors: Voices from the Epidemic*. Oxford: Oxford University Press.

Bayles, Martha. 1999. "Imus, Oprah and the Literary Elite." *New York Times Book Review*, 29 August, 35.

Bayley, John. 1998. *Iris: A Memoir of Iris Murdoch*. London: Duckworth.

———. 2000a. *Iris and Her Friends: A Memoir of Memory and Desire*. New York and London: W. W. Norton and Co.

———. 2000b. "Iris Would Have Warmly Approved." *Daily Telegraph*, 15 January, 21.

———. 2000c. "The Quartet That Became a Couple." *Daily Telegraph*, 15 June, 22.

———. 2001. *Widower's House: A Study in Bereavement, or How Margot and Mella Forced Me to Flee My Home*. New York: W. W. Norton and Co.

Benhabib, Seyla. 1995. "Feminism and Postmodernism." In *Feminist Contentions: A Philosophical Exchange*, 17–34, by Benhabib et al. New York and London: Routledge.

Bersani, Leo. 1986. *The Freudian Body: Psychoanalysis and Art*. New York: Columbia University Press.

Blanchot, Maurice. 1995. *The Writing of the Disaster*. Trans. Ann Smock. Lincoln and London: University of Nebraska Press.

Boehmer, Ulrike. 2000. *The Personal and the Political: Women's Activism in Response to the Breast Cancer and AIDS Epidemics*. Albany: State University of New York Press.

Bollas, Christopher. 1995. *Cracking Up: The Work of Unconscious Experience*. New York: Hill and Wang.

Bordo, Susan. 1990. "Reading the Slender Body." In *Body/Politics: Women and the Discourses of Science*, ed. Mary Jacobus, Evelyn Fox Keller, and Sally Shuttleworth, 83–112. New York and London: Routledge.

———. 1993. *Unbearable Weight: Feminism, Western Culture, and the Body*. Berkeley: University of California Press.

Boseley, Sarah. 1999. "300,000 Payout Over Writer's Cancer Death." *Guardian*, 12 November, 6.

Boston Women's Health Collective. 1973. *Our Bodies, Ourselves*. New York: Simon and Schuster.

Bourdieu, Pierre. 1984. *Distinction: A Social Critique of the Judgment of Taste*. Trans. Richard Nice. Cambridge, Mass.: Harvard University Press.

Braidotti, Rosi. 1994. *Nomadic Subjects: Embodiment and Sexual Difference in Contemporary Feminist Theory*. New York: Columbia University Press.

Brennan, Mary. 1995. "No-one Gets Out of Here Alive." *Herald* (Glasgow), 5 August, 16.

Brodribb, Somer. 1992. *Nothing Mat(t)ers: A Feminist Critique of Postmodernism*. Melbourne: Spinifex Press.

Brody, Howard. 1987. *Stories of Sickness*. New Haven, Conn.: Yale University Press.

Bronfen, Elisabeth. 1992. *Over Her Dead Body: Death, Femininity, and the Aesthetic*. New York: Routledge.

Butler, Judith. 1989. "Sexual Ideology and Phenomenological Description: A Feminist Critique of Merleau-Ponty's *Phenomenology of Perception*." In *The Thinking Muse: Feminism and Modern French Philosophy*, ed. Jeffner Allen and Iris Marion Young, 85–100. Bloomington: Indiana University Press.

———. 1990. *Gender Trouble: Feminism and the Subversion of Identity*. New York and London: Routledge.

———. 1993a. *Bodies That Matter: On the Discursive Limits of "Sex."* New York and London: Routledge.

———. 1993b. "Imitation and Gender Insubordination." In *The Lesbian and Gay Studies Reader*, ed. Henry Abelove, Michele Aina Barale, and David M. Halperin, 307–20. New York and London: Routledge.

———. 1995. "Contingent Foundations." In *Feminist Contentions: A Philosophical Exchange*, 35–57, by Benhabib et al. New York and London: Routledge.

———. 1997a. *Excitable Speech: A Politics of the Performative*. New York and London: Routledge.

———. 1997b. *The Psychic Life of Power: Theories in Subjection*. Stanford: Stanford University Press.

———. 2000. *Antigone's Claim: Kinship between Life and Death*. New York: Columbia University Press.

Butler, Sandra, and Barbara Rosenblum. 1991. *Cancer in Two Voices*. San Francisco: Spinsters Book Company.

Cady, Joseph. 1993. "Immersive and Counterimmersive Writing about AIDS: The Achievement of Paul Monette's *Love Alone*." In *Writing AIDS: Gay Literature, Language, and Analysis*, ed. Timothy F. Murphy and Suzanne Poirier, 244–64. New York: Columbia University Press.

Callahan, John. 1989. *Don't Worry, He Won't Get Far on Foot: The Autobiography of a Dangerous Man*. New York: Morrow.

Campo, Rafael. 1997. *The Poetry of Healing: A Doctor's Education in Empathy, Identity, and Desire*. New York and London: W. W. Norton and Co.

Canguilhem, Georges. 1991. *The Normal and the Pathological*. Trans. Carolyn R. Fawcett. New York: Zone Books.

Carr, David. 1986. *Time, Narrative, and History*. Bloomington: Indiana University Press.

Carter, Erica, and Simon Watney, eds. 1989. *Taking Liberties*. London: Serpent's Tail.

Caruth, Cathy, ed. 1995. *Trauma: Explorations in Memory*. Baltimore: Johns Hopkins University Press.

———. 1996. *Unclaimed Experience: Trauma, Narrative, and History*. Baltimore: Johns Hopkins University Press.

Caruth, Cathy, and Thomas Keenan. 1995. "'The AIDS Crisis Is Not Over': A Conversation with Gregg Bordowitz, Douglas Crimp, and Laura Pinsky." In *Trauma: Explorations in Memory*, ed. Cathy Caruth, 256–71. Baltimore: Johns Hopkins University Press.

Certeau, Michel de. 1984. *The Practice of Everyday Life*. Trans. Steven Rendall. Berkeley: University of California Press.

Chambers, Ross. 1998. *Facing It: AIDS Diaries and the Death of the Author*. Ann Arbor: University of Michigan Press.

Chambers, Tod. 1999. *The Fiction of Bioethics: Cases as Literary Texts*. London and New York: Routledge.

Chandler, Katherine R., and Melissa A. Goldthwaite, eds. 2003. *Surveying the Literary Landscapes of Terry Tempest Williams: New Critical Essays*. Salt Lake City: University of Utah Press.

Charon, Rita, and Martha Montello, eds. 2002. *Stories Matter: The Role of Narrative in Medical Ethics*. New York: Routledge.

Clarke, Brock. 1999. "Why Memoir Isn't Always Art." *Chronicle of Higher Education*, 29 October, B9.

Clarke, Bruce, and Wendell Aycock, eds. 1990. *The Body and the Text: Comparative Essays in Literature and Medicine*. Lubbock: Texas Tech University Press.

Clum, John M. 1993 "'And Once I Had It All': AIDS Narratives and Memories of an American Dream." In *Writing AIDS: Gay Literature, Language, and Analysis*, ed. Timothy F. Murphy and Suzanne Poirier, 200–224. New York: Columbia University Press.

Coburn, Judith. 2000. "Off the Pill." In *Dear Sisters: Dispatches from the Women's Liberation Movement*, ed. Rosalyn Baxandall and Linda Gordon, 136–39. New York: Basic Books.

Colbourn, Carole. 1996. "Refugees in a Strange Country." In *Cancer: Through the Eyes of Ten Women*, ed. Patricia Duncker and Vicky Wilson. London: Pandora.

Cole, Luke W., and Sheila R. Foster. 2001. *From the Ground Up: Environmental Racism and the Rise of the Environmental Justice Movement*. New York: New York University Press.

Comaroff, Jean, and John L. Comaroff, eds. 2000. "Millennial Capitalism and the Culture of Neoliberalism." Special Issue of *Public Culture* 12, no. 2.

Couser, G. Thomas. 1997. *Recovering Bodies: Illness, Disability, and Life Writing*. Madison: University of Wisconsin Press.

Coward, Rosalind. 1989. *The Whole Truth: The Myth of Alternative Health*. London and Boston: Faber and Faber.

Craddock, Susan. 2000. *City of Plagues: Disease, Poverty, and Deviance in San Francisco*. Minneapolis: University of Minnesota Press.

Crawford, Robert. 1984. "A Cultural Account of 'Health': Control, Release, and the Social Body." In *Issues in the Political Economy of Health Care*, ed. John B. McKinlay, 60–103. New York and London: Tavistock Publications.

Crimp, Douglas, ed. 1988. *AIDS: Cultural Analysis/Cultural Activism*. Cambridge, Mass.: MIT Press.

———. 1990. "Mourning and Militancy." In *Out There: Marginalization and Contemporary Culture*, ed. Russell Ferguson, Martha Gever, Trinh T. Minh-ha, and Cornel West, 233–45. New York: The New Museum of Contemporary Art.

Croce, Arlene. 1994/1995. "Discussing the Undiscussable." *The New Yorker*, 26 December–2 January, 54–60.

Cushman, Philip. 1995. *Constructing the Self, Constructing America: A Cultural History of Psychotherapy*. Reading, Mass.: Addison-Wesley.

Cvetkovich, Ann. 2003. *An Archive of Feelings: Trauma, Sexuality, and Lesbian Public Cultures*. Durham, N.C.: Duke University Press.

Daniel, Rosy. 2000. *Living with Cancer*. London: Constable and Robinson.

Davis, Angela Y. 1983. *Women, Race, and Class*. New York: Vintage Books.

Davis, Flora. 1999. *Moving the Mountain: The Women's Movement in America since 1960*. Urbana and Chicago: University of Illinois Press.

Davis, Lennard. 1995. *Enforcing Normalcy: Disability, Deafness, and the Body*. London and New York: Verso.

Delany, Samuel R. 2004. *The Motion of Light in Water: Sex and Science Fiction Writing in the East Village*. Rev. ed. Minneapolis: University of Minnesota Press.

Deleuze, Gilles. 1990. *Negotiations, 1972–1990*. New York: Columbia University Press.

Deleuze, Gilles, and Félix Guattari. 1986. *Nomadology: The War Machine*. Trans. Brian Massumi. New York: Semiotexte.

———. 1987. *A Thousand Plateaus: Capitalism and Schizophrenia*. Trans. Brian Massumi. Minneapolis: University of Minnesota Press.

de Man, Paul. 1984. *The Rhetoric of Romanticism*. New York: Columbia University Press.

Derrida, Jacques. 1982. *Margins of Philosophy*. Trans. by Alan Bass. Chicago: University of Chicago Press.

Diamond, John. 1998. *C: Because Cowards Get Cancer Too. . . .* London: Vermilion.

Diedrich, Lisa. 2001. "Breaking Down: A Phenomenology of Disability." *Literature and Medicine* 20, no. 2: 209–30.

———. 2003. "'Meanwhile the world goes on': Terry Tempest Williams and an Ethics of Attention." In *Surveying the Literary Landscapes of Terry Tempest Williams: New Critical Essays*, ed. Katherine R. Chandler and Melissa A. Goldthwaite, 211–28. Salt Lake City: University of Utah Press.

———. 2005. "Introduction: Genealogies of Disability: Historical Emergences and Everyday Enactments," *Cultural Studies*, 19, no. 6: 649–66.

———. 2007 (forthcoming). "Gathering Evidence of Ghosts: W. G. Sebald's Practices of Witnessing. In *Searching for Sebald*, ed. Lise Patt. Los Angeles: ICI Press.

Diprose, Rosalyn. 1994. *The Bodies of Women: Ethics, Embodiment and Sexual Difference*. London and New York: Routledge.

Donoghue, Denis. 1978. "Disease Should Be Itself." *New York Times Book Review*, 16 July, 9 and 27.

Dormandy, Thomas. 2000. *The White Death: A History of Tuberculosis*. New York: New York University Press.

Doty, Mark. 1996. *Heaven's Coast: A Memoir*. New York: HarperCollins.

Dreifus, Claudia, ed. 1977. *Seizing Our Bodies: The Politics of Women's Health*. New York: Vintage.

Dresser, Rebecca. 1996. "What Bioethics Can Learn from the Women's Health Movement." In *Feminism and Bioethics: Beyond Reproduction*, ed. Susan M. Wolf, 144–59. New York: Oxford University Press.

Driver, Felix. 1994. "Bodies in Space: Foucault's Account of Disciplinary Power." In *Reassessing Foucault: Power, Medicine, and the Body*, ed. Colin Jones and Roy Porter, 113–31. London and New York: Routledge.

Dubos, René, and Jean Dubos. 1952. *The White Plague: Tuberculosis, Man, and Society*. New Brunswick, N.J.: Rutgers University Press.

Duggan, Lisa. 2003. *The Twilight of Equality? Neoliberalism, Cultural Politics, and the Attack on Democracy*. Boston: Beacon Press.

Eagleton, Terry. 1983. *Literary Theory: An Introduction*. Minneapolis: University of Minnesota Press.

Eakin, Paul John. 1999. *How Our Lives Become Stories: Making Selves*. Ithaca, N.Y.: Cornell University Press.

Edson, Margaret. 1999. *Wit*. New York: Faber and Faber.

Egan, Susanna. 1999. *Mirror Talk: Genres of Crisis in Contemporary Autobiography*. Chapel Hill: University of North Carolina Press.

Ehrenreich, Barbara. 2001. "Welcome to Cancerland." *Harper's Magazine*, November, 43–53.

Ehrenreich, Barbara, and Deirdre English. 1978. *For Her Own Good: 150 Years of the Experts' Advice to Women*. New York: Anchor.

Eisenstein, Zillah. 2001. *Manmade Breast Cancers*. Ithaca, N.Y.: Cornell University Press.

Elkins, James. 1999. *Pictures of the Body: Pain and Metamorphosis*. Stanford: Stanford University Press.

Ellison, Michael. 1999. "Dying Columnist Sacked for Keeping Secret." *Guardian*, 13 May, 15.

Engel, George. 1977. "The Need for a New Medical Model: A Challenge for Biomedicine." *Science* 196: 129–36.

Epstein, Julia. 1995. *Altered Conditions: Disease, Medicine, and Storytelling*. New York and London: Routledge.

Epstein, Leslie. 2000. "Odessa in the Bronx." *New York Times Book Review*, 9 July, 13.

Epstein, Steven. 1996. *Impure Science: AIDS, Activism, and the Politics of Knowledge*. Berkeley: University of California Press.

Farmer, Paul. 1999. *Infections and Inequalities: The Modern Plagues*. Berkeley: University of California Press.

———. 2003. *Pathologies of Power: Health, Human Rights, and the New War on the Poor*. Berkeley: University of California Press.

Feldberg, Georgina D. 1995. *Disease and Class: Tuberculosis and the Shaping of Modern North American Society*. New Brunswick, N.J.: Rutgers University Press.

Felman, Shoshana, and Dori Laub. 1992. *Testimony: Crisis of Witnessing in Literature, Psychoanalysis, and History*. New York and London: Routledge.

Ferraro, Susan. 1993. "'You Can't Look Away Anymore': The Anguished Politics of Breast Cancer." *New York Times Magazine*, 15 April, 24–25, 58, 61–62.

Fisher, Gary. 1996. *Gary in Your Pocket: Stories and Notebooks of Gary Fisher*. Ed. Eve Kosofsky Sedgwick. Durham, N.C.: Duke University Press.

Fleck, Ludwik. 1979 [1935]. *Genesis and Development of a Scientific Fact*. Trans. Fred Bradley and Thaddeus J. Trenn. Chicago: University of Chicago Press.

Foucault, Michel. 1972. *The Archaeology of Knowledge and the Discourse on Language*. Trans. A. M. Sheridan Smith. New York: Pantheon.

———. 1973. *The Birth of the Clinic: An Archeology of Medical Perception*. Trans. A. M. Sheridan Smith. New York: Vintage.

———. 1977a. *Discipline and Punish: The Birth of the Prison*. Trans. Alan Sheridan. New York: Vintage.

———. 1977b. *Language, Counter-Memory, Practice: Selected Essays and Interviews*. Trans. Donald F. Bouchard and Sherry Simon. Ithaca, N.Y.: Cornell University Press.

———. 1978. *An Introduction*. Vol. 1 of *The History of Sexuality*. Trans. Robert Hurley. New York: Vintage.

———. 1980. *Power/Knowledge: Selected Interviews and Other Writings 1972–1977*. Trans. Colin Gordon et. al. New York: Pantheon.

———. 1985. *The Use of Pleasure*. Vol. 2 of *The History of Sexuality*. Trans. Robert Hurley. New York: Vintage.

———. 1986a. *The Care of the Self*. Vol. 3 of *The History of Sexuality*. Trans. Robert Hurley. New York: Vintage.

———. 1986b. "Maurice Blanchot: The Thought from Outside." In *Foucault/Blanchot*, trans. Brian Massumi, 9–58. Seattle: Zone Books.

———. 1991. Introd. to *The Normal and the Pathological*, by Georges Canguilhem, trans. Carolyn R. Fawcett, 7–24. New York: Zone Books.

———. 1997. *Ethics: Subjectivity and Truth*. Vol. 1 of *Essential Works of Foucault, 1954–1984*. Ed. Paul Rabinow. Trans. Robert Hurley and Others. New York: The New Press.

———. 1998. *Aesthetics, Method, and Epistemology*. Vol. 2 of *Essential Works of Foucault, 1954–1984*. Ed. Paul Rabinow. Trans. Robert Hurley and Others. New York: The New Press.

———. 2003. *"Society Must Be Defended": Lectures at the Collège de France, 1975–76*, ed. Mauro Bertani and Alessandro Fontana, trans. David Macey. New York: Picador.

Frank, Arthur W. 1991. *At the Will of the Body: Reflections on Illness*. Boston: Houghton Mifflin.

———. 1995. *The Wounded Storyteller: Body, Illness, and Ethics*. Chicago: University of Chicago Press.

———. 2004. *The Renewal of Generosity: Illness, Medicine, and How to Live*. Chicago: University of Chicago Press.

Frankfort, Ellen. 1972. *Vaginal Politics*. New York: Quadrangle Books.

———. 1977. "Vaginal Politics." In *Seizing Our Bodies: The Politics of Women's Health*, ed. Claudia Dreifus, 263–70. New York: Vintage.

Franklin, Benjamin. 1993 [1791]. *Autobiography and Other Writings*. Ed. Ormond Seavey. New York: Oxford University Press.

Fraser, Nancy, and Linda Nicholson. 1990. "Social Criticism without Philosophy: An Encounter between Feminism and Postmodernism." In *Feminism/Postmodernism*, ed. Linda J. Nicholson, 19–38. New York and London: Routledge.

Freud, Sigmund. 1953 [1919]. "The Uncanny." In *Collected Papers, Volume IV*, ed. Ernest Jones, trans. Joan Riviere, 368–407. London: The Hogarth Press.

———. 1961 [1920]. *Beyond the Pleasure Principle*. Trans. James Strachey. New York: W. W. Norton and Co.

———. 1961 [1930]. *Civilization and Its Discontents*. Trans. James Strachey. New York: W. W. Norton and Co.

———. 1965 [1900]. *The Interpretation of Dreams*. Trans. and ed. James Strachey. New York: Avon Books.

———. 1967 [1939]. *Moses and Monotheism*. Trans. Katherine Jones. New York: Vintage.

Friedman, Richard. 2005. "Learning Words They Rarely Teach in Medical School: 'I'm Sorry.'" *New York Times*, 26 July, 5.

Fruchter, Rachel Gillett, et al. 1977. "The Women's Health Movement: Where Are We Now?" In *Seizing Our Bodies: The Politics of Women's Health*, ed. Claudia Dreifus, 271–78. New York: Vintage.

Garrett, Laurie. 1994. *The Coming Plague: Newly Emerging Diseases in a World Out of Balance*. New York: Farrar, Straus and Giroux.

Gawande, Atul. 2002. *Complications: A Surgeon's Notes on an Imperfect Science*. New York: Picador.

Geefe, Richard. 1999a. "Time to Go." *Observer*, 30 May, 2.

———. 1999b. "A guy tries to kill himself and you organize a picnic?" *Observer*, 6 June, 4.

———. 1999c. "There I was surrounded by famous and beautiful people while Bowie fiddled alone with his sarong." *Observer*, 13 June, 4.

———. 1999d. "Richard Geefe: The Complete Works: The Suicide Journalist." *Observer*, 4 July, 9 and 12.

Geertz, Clifford. 1978. "Stir Crazy: A Review of *Discipline and Punish*." *New York Review of Books*, 26 January, 3–6.

Gilbert, Jenny. 1998. "Still Here, Still Stirring." *Independent* (London), 15 March, 16.

Gilbey, Emma. 1995. "Death Takes Centre Stage." *Sunday Telegraph* (London), 15 January, 24.

Gilligan, Carol. 1982. *In a Different Voice: Psychological Theory and Women's Development*. Cambridge, Mass.: Harvard University Press.

Gilman, Sander L. 1988. *Disease and Representation: Images of Illness from Madness to AIDS*. Ithaca, N.Y.: Cornell University Press.

———. 1993. "Plague in Germany, 1939/1989: Cultural Images of Race, Space, and Disease." In *Writing AIDS: Gay Literature, Language, and Analysis*, ed. Timothy F. Murphy and Suzanne Poirier, 54–82. New York: Columbia University Press.

Gordon, Avery. 1997. *Ghostly Matters: Haunting and the Sociological Imagination*. Minneapolis: University of Minnesota Press.

Grant, Linda. 1999. "True Confessions." *Guardian*, 29 June, 13.

Grosz, Elizabeth. 1994. *Volatile Bodies: Toward a Corporeal Feminism*. Bloomington and Indianapolis: Indiana University Press.

———. 1999. "Becoming [. . .]: An Introduction." In *Becomings: Explorations in Time, Memory, and Futures*, ed. Elizabeth Grosz, 1–11. Ithaca, N.Y.: Cornell University Press.

Grover, Jan Zita. 1989. "Constitutional Symptoms." In *Taking Liberties*, ed. Erica Carter and Simon Watney, 147–59. London: Serpent's Tail.

Gunther, John. 1949. *Death Be Not Proud: A Memoir*. New York: Harper and Row.

Gusdorf, Georges. 1980. "Conditions and Limits of Autobiography." In *Autobiography: Essays Theoretical and Critical*, ed. and trans. James Olney, 28–48. Princeton, N.J.: Princeton University Press.

Hammond, Michael, Jane Howarth, and Russell Kent. 1991. *Understanding Phenomenology*. Oxford: Basil Blackwell.

Hankivsky, Olena. 2004. *Social Policy and the Ethic of Care*. Vancouver: University of British Columbia Press.

Haraway, Donna. 1991. *Simians, Cyborgs, and Women: The Reinvention of Nature*. New York: Routledge.

Harrison, Kathryn. 1997. *The Kiss*. New York: Random House.

Hawkins, Anne Hunsaker. 1993. *Reconstructing Illness: Studies in Pathography*. West Lafayette, Ind.: Purdue University Press.

Head, Henry. 1920. *Studies in Neurology: Vol. 1*. London: Hodder and Stoughton.

Heidegger, Martin. 1962. *Being and Time*. Trans. John Macquarrie and Edward Robinson. San Francisco: HarperCollins.

Hesford, Victoria. 2001. "Becoming Feminism: The Emergence of the Feminist-as-Lesbian Figure and the Forgetting of the Women's Movement." Ph.D. diss., Emory University.

Hilfiker, David. 1985. *Healing the Wounds: A Doctor Looks at His Work*. New York: Pantheon.

Hockenberry, John. 1995. *Moving Violations: War Zones, Wheelchairs, and Declarations of Independence*. New York: Hyperion.

Howard, John. 1999. *Men Like That: A Southern Queer History*. Chicago: University of Chicago Press.

Hunter, Kathryn Montgomery. 1991. *Doctors' Stories: The Narrative Structure of Medical Knowledge*. Princeton, N.J.: Princeton University Press.

Husserl, Edmund. 1960. *Cartesian Meditations: An Introduction to Phenomenology*. Trans. Dorion Cairns. The Hague: Martinus Nijhoff.

Institute of Medicine. 2000. *To Err Is Human: Building a Safer Health System*. Washington, D.C.: National Academy Press.

Jacobs, Carol. 1994. *Telling Time: Lévi-Strauss, Ford, Lessing, Benjamin, de Man, Wordsworth, Rilke*. Baltimore: Johns Hopkins University Press.

Jameson, Frederic. 1972. *The Prison-House of Language: A Critical Account of Structuralism and Russian Formalism*. Princeton, N. J.: Princeton University Press.

Jarman, Derek. 1992. *At Your Own Risk: A Saint's Testament*, ed. Michael Christie. London: Vintage.

Jetter, Alexis. 2000. "Breast Cancer in Blacks Spurs Hunt for Answers." *New York Times*, 22 February, D5.

Johnson, Daniel. 2000. "Don't Write Yourself Off Too Soon, Martin." *Daily Telegraph*, 2 May, 20.

Johnson, Mark. 1987. *The Body in the Mind: The Bodily Basis of Meaning, Imagination, and Reason*. Chicago: University of Chicago Press.

Kaplan, Alice. 1993. *French Lessons: A Memoir*. Chicago: University of Chicago Press.

———. 1994. "On Language Memoir." In *Displacements: Cultural Identities in Question*, ed. Angelika Bammer, 59–70. Bloomington and Indianapolis: Indiana University Press.

Karr, Mary. 1995. *The Liars' Club: A Memoir*. New York: Viking.

Katz, Jay. 1984. *The Silent World of Doctor and Patient*. New York: Free Press.

Kayal, Philip M. 1993. *Bearing Witness: Gay Men's Health Crisis and the Politics of AIDS*. Boulder, Colo.: Westview Press.

Kempe, Margery. 2001 [ca. 1373]. *The Book of Margery Kempe*. Trans. Lynn Staley. New York: W. W. Norton and Co.

Kestenbaum, Victor, ed. 1982. *The Humanity of the Ill: Phenomenological Perspectives*. Knoxville: University of Tennessee Press.

Kimmich, Allison. 1998. "Writing the Body: From Abject to Subject." *A/B: Auto/Biography* 13, no. 2: 223–34.

Kincaid, Jamaica. 1997. *My Brother*. New York: Farrar, Straus and Giroux.

Kittay, Eva Feder, and Ellen K. Feder. 2003. *The Subject of Care: Feminist Perspectives on Dependency*. Lanham, Md.: Rowman and Littlefield.

Kleinman, Arthur. 1988. *The Illness Narratives: Suffering, Healing, and the Human Condition*. New York: Basic Books.

Kleinman, Arthur, Veena Das, and Margaret M. Lock, eds. 1997. *Social Suffering*. Berkeley: University of California Press.

Kramer, Larry. 1985. *The Normal Heart*. New York: Methuen.

———. 1994. *Reports from the Holocaust: The Story of an AIDS Activist*. 2nd ed. New York: St. Martin's Press.

Kuczynski, Alex. 1999. "Fired Reporter Says She Lied About AIDS, Too." *New York Times*, 15 May, Section A, 18.

Kushner, Rose. 1975. *Why Me? What Every Woman Should Know About Breast Cancer to Save Her Life*. New York: New American Library.

———. 1977. "The Politics of Breast Cancer." In *Seizing Our Bodies: The Politics of Women's Health*, ed. Claudia Dreifus, 186–94. New York: Vintage.

Kushner, Tony. 1995. "Letters." *The New Yorker*, 30 January, 11.

Lacan, Jacques. 1977. *Ecrits: A Selection*. Trans. Alan Sheridan. New York: W. W. Norton and Co.

———. 1981. *The Four Fundamental Concepts of Psychoanalysis*. Trans. Alan Sheridan. Ed. Jacques-Alain Miller. New York: W. W. Norton and Co.

———. 1992. *The Ethics of Psychoanalysis, 1959–1960: Seminar VII*. Trans. Dennis Porter. Ed. Jacques-Alain Miller. New York: W. W. Norton and Co.

LaCapra, Dominick. 1994. *Representing the Holocaust: History, Theory, Trauma*. Ithaca, N.Y.: Cornell University Press.

La Motte, Ellen N. 1915. *The Tuberculosis Nurse*. New York: Garland, 1985.

Langer, Lawrence. 1991. *Holocaust Testimonies: The Ruins of Memory*. New Haven, Conn.: Yale University Press.

Latour, Bruno. 1993. *We Have Never Been Modern*. Trans. Catherine Porter. Cambridge, Mass.: Harvard University Press.

Latour, Bruno, and Steve Woolgar. 1986. *Laboratory Life: The Construction of Scientific Facts*. 2d ed. Princeton, N.J.: Princeton University Press.

Laub, Dori. 1992. "Bearing Witness, or the Vicissitudes of Listening." In *Testimony: Crises of Witnessing in Literature, Psychoanalysis, and History*, ed. Shoshana Felman and Dori Laub, 57–74. New York and London: Routledge.

———. 1995. "Truth and Testimony: The Process and the Struggle." In Trauma: Explorations in Memory, ed. Cathy Caruth, 61–75. Baltimore: Johns Hopkins University Press.

Law, John. 1994. *Organizing Modernity*. Oxford: Blackwell.

Law, John, and Annemarie Mol. 2002. "Complexities: An Introduction." In *Complexities: Social Studies of Knowledge Practices*, ed. John Law and Annemarie Mol, 1–22. Durham, N.C.: Duke University Press.

Leo, John. 1995. "The Backlash against Victim Art." *U.S. News and World Report*, 16 January, 22.

Leonard, John. 1978. "Books of the Times: *Illness as Metaphor*." *New York Times*, 1 June, C19.

Leopold, Ellen. 2000. *A Darker Ribbon: Breast Cancer, Women, and Their Doctors in the Twentieth Century*. Boston: Beacon Press.

Lerner, Barron. 2001. *The Breast Cancer Wars: Hope, Fear, and the Pursuit of a Cure in Twentieth-Century America*. Oxford: Oxford University Press.

Levinas, Emmanuel. 1983. "Beyond Intentionality." In *Philosophy in France Today*, ed. A. Montefiore, 100–115. Cambridge: Cambridge University Press.

———. 1989. *The Levinas Reader*. Ed. Seán Hand. Oxford: Blackwell.

Lopez, Diego, and George S. Getzel. 1987. "Strategies for Volunteers Caring for Persons with AIDS." *Social Casework: The Journal of Contemporary Social Work* (January): 47–53.

Lorde, Audre. 1980. *The Cancer Journals*. San Francisco: Aunt Lute Books.

———. 1982. *Zami: A New Spelling of My Name*. Freedom, Calif.: The Crossing Press.

———. 1984. *Sister Outsider: Essays and Speeches*. Trumansburg, N.Y.: Crossing Press.

———. 1988. *A Burst of Light: Essays*. Ithaca, N.Y.: Firebrand Books.

———. 1993. *The Marvelous Arithmetics of Distance: Poems 1987–1992*. New York and London: W. W. Norton and Co.

———. 1993 [1978]. "The Uses of the Erotic: The Erotic as Power." In *The Lesbian and Gay Studies Reader*, ed. Henry Abelove, Michele Aina Barale, and David M. Halperin, 339–43. New York and London: Routledge.

———. 1997a. *The Cancer Journals: Special Edition*. San Francisco: Aunt Lute Books.

———. 1997b. *The Collected Poems of Audre Lorde*. New York: W. W. Norton and Co.

Love, Dr. Susan, with Karen Lindsey. 1990. *Dr. Susan Love's Breast Book*. Reading, Mass.: Addison-Wesley.

Lyotard, Jean-François. 1984. *The Postmodern Condition: A Report on Knowledge*. Trans. Geoff Bennington and Brian Massumi. Minneapolis: University of Minnesota Press.

———. 1988. *The Differend: Phrases in Dispute*. Trans. Georges Van Den Abbeele. Minneapolis: University of Minnesota Press.

MacDonald, Betty Bard. 1945. *The Egg and I*. Philadelphia and New York: J. B. Lippincott.

———. 1948. *The Plague and I*. Philadelphia and New York: J. B. Lippincott.

MacGregor, Jeff. 1998. "Inner Peace, Empowerment and Host Worship." *New York Times*, 25 October, Arts Section, 30.

Mackrell, Judith. 1995. "A Dance About AIDS?" *Guardian* (London), 28 August, T9.

Mairs, Nancy. 1989. *Remembering the Bone House: An Erotics of Place and Space*. New York: Harper and Row.

———. 1994. *Voice Lessons: On Becoming a (Woman) Writer*. Boston: Beacon Press.

———. 1996. *Waist-High in the World: A Life Among the Disabled*. Boston: Beacon Press.

———. 1997. "Introduction." In *Recovering Bodies: Illness, Disability, and Life Writing*, by G. Thomas Couser. Madison: University of Wisconsin Press.

Mars-Jones, Adam. 1995. "Victims Don't Make Art." *Independent* (London), 15 March, 24.

Massumi, Brian. 1987. "Translator's Foreword: Pleasures of Philosophy." In *A Thousand Plateaus: Capitalism and Schizophrenia*, by Gilles Deleuze and Félix Guattari. Trans. Brian Massumi, ix–xix. Minneapolis: University of Minnesota Press.

Mauss, Marcel. 1990. *The Gift*. Trans. W. D. Halls. New York and London: W. W. Norton and Co.

Max, D. T. 1999. "The Oprah Effect." *New York Times Magazine*, 26 December, 36–37; 40–41.

Merleau-Ponty, Maurice. 1962. *Phenomenology of Perception*. Trans. Colin Smith. London and New York: Routledge.

Miller, D. A. 1993. "Sontag's Urbanity." In *The Lesbian and Gay Studies Reader*, ed. Henry Abelove, Michele Aina Barale, and David M. Halperin, 212–20. New York and London: Routledge.

Miller, Nancy K. 1996. *Bequest and Betrayal: Memoirs of a Parent's Death*. New York and Oxford: Oxford University Press.

Minnie the Minx. 1999. "Media News: Diary." *Daily Telegraph*, 11 June, 21.

Mohanty, Chandra. 1994. "Under Western Eyes: Feminist Scholarship and Colonial Discourses." In *Colonial Discourse and Post-Colonial Theory: A Reader*, ed. Patrick Williams and Laura Chrisman, 196–220. New York: Columbia University Press.

Mol, Annemarie. 2002. *The Body Multiple: Ontology in Medical Practice*. Durham, N.C.: Duke University Press.

Monette, Paul. 1988a. *Borrowed Time: An AIDS Memoir*. San Diego, Calif.: Harcourt Brace Jovanovich.

———. 1988b. *Love Alone: Eighteen Elegies for Rog*. New York: St. Martin's Press.

———. 1992. *Becoming a Man: Half a Life Story*. New York: Harcourt Brace Jovanovich.

———. 1994. *Last Watch of the Night*. New York: Harcourt Brace & Company.

———. 1997. *Sanctuary: A Tale of Life in the Woods*. New York: Scribner.

Moore, Oscar. 1996. *PWA: Looking AIDS in the Face*. London: Picador.

Morgen, Sandra. 2002. *Into Our Own Hands: The Women's Health Movement in the United States, 1969–1990*. New Brunswick, N.J.: Rutgers University Press.

Morris, David. 1998. *Illness and Culture in the Postmodern Age*. Berkeley: University of California Press.

Morris, Meaghan. 1998. *Too Soon Too Late: History in Popular Culture*. Bloomington and Indianapolis: Indiana University Press.

Murdoch, Iris. 1976. *The Severed Head*. New York: Penguin.

———. 1978. *The Sea, the Sea*. New York: Viking.

———. 1992. *Metaphysics as a Guide to Morals*. London and New York: Penguin.

———. 1995. *Jackson's Dilemma*. London: Chatto and Windus.

Murphy, Robert. 1987. *The Body Silent*. New York: Henry Holt and Co.

Murphy, Timothy F. 1993. "Testimony." In *Writing AIDS: Gay Literature, Language, and Analysis*, ed. Timothy F. Murphy and Suzanne Poirier, 306–20. New York: Columbia University Press.

Murphy, Timothy F., and Suzanne Poirier, eds. 1993. *Writing AIDS: Gay Literature, Language, and Analysis*. New York: Columbia University Press.

Nelson, Hilde Lindemann. 2001. *Damaged Identities, Narrative Repair*. Ithaca, N.Y.: Cornell University Press.

New Yorker. 1995. "Who's the Victim? Dissenting Voices Answer Arlene Croce's Critique of Victim Art." *The New Yorker*, 30 January, 10–13.

New York Times. 1978. "Susan Sontag Found Crisis of Cancer Added a Fierce Intensity to Life." *New York Times*, 30 January, A16.

———. 1986. "Words That Heal: Doctors Talk About Their Writing." *New York Times Book Review*, 23 March, 33.

Noddings, Nel. 1984. *Caring: A Feminist Approach to Ethics and Moral Education*. Berkeley: University of California Press.

Nuland, Sherwin B. 1994. *How We Die: Reflections on Life's Final Chapter*. New York: Alfred A. Knopf.

Oliver, Kelly. 1997. *Family Values: Subjects between Nature and Culture*. New York: Routledge.

———. 2001. *Witnessing: Beyond Recognition*. Minneapolis: University of Minnesota Press.

———. 2002. "Psychic Space and Social Melancholy." In *Between the Psyche and the Social: Psychoanalytic Social Theory*, ed. Kelly Oliver and Steve Edwin, 49–65. Lanham, Md.: Rowman and Littlefield.

Olney, James. 1980. "Autobiography and the Cultural Moment: A Thematic, Historical, and Bibliographical Introduction." In *Autobiography: Essays Theoretical and Critical*, ed. James Olney, 3–27. Princeton, N.J.: Princeton University Press.

Ott, Katherine. 1996. *Fevered Lives: Tuberculosis in American Culture since 1870*. Cambridge, Mass.: Harvard University Press.

Parini, Jay. 1998. "Delving into the World of Dreams by Blending Fact and Fiction." *Chronicle of Higher Education*, 27 February, A40.

Parker, Andrew, and Eve Kosofsky Sedgwick. 1995. "Introduction." In *Performativity and Performance*, ed. Andrew Parker and Eve Kosofsky Sedgwick, 1–18. New York: Routledge.

Patton, Cindy. 1989. "The AIDS Industry: Construction of 'Victims,' 'Volunteers,' and 'Experts.'" In *Taking Liberties*, ed. Erica Carter and Simon Watney, 113–25. London: Serpent's Tail.

———. 1990. *Inventing AIDS*. New York and London: Routledge.

Pearl, Monica B. 2003. "Eve Sedgwick's Melancholic 'White Glasses.'" *Textual Practice* 17, no. 1: 61–80.

Peck, Janice. 1995. "TV Talk Shows as Therapeutic Discourse: The Ideological Labor of the Televised Talking Cure." *Communication Theory* 5, no. 1 (February): 58–81.

Phillips, Adam. 2000. *Darwin's Worms*. New York: Basic Books.

Picardie, Justine. 2000. "If I Dream I Have You." *Granta* 71 (Autumn): 163–84.

Picardie, Ruth. 1998. *Before I Say Goodbye*. London: Penguin.

Pierce, Charles P. 2000. *Hard to Forget: An Alzheimer's Story*. New York: Random House.

Poirier, Suzanne. 1993. "On Writing AIDS: Introduction." In *Writing AIDS: Gay Literature, Language, and Analysis*, ed. Timothy F. Murphy and Suzanne Poirier, 1–8. New York: Columbia University Press.

Poynter, Chris. 1999. "Reporter's AIDS Story Also Untrue." *Courier-Journal* (Louisville, Kentucky), 15 May, 15.

Price, Catherine. 1992. Review of *Madonna Swan: A Lakota Woman's Story*. *The Journal of American History* 79, no. 3: 1223–24.

Price, Janet, and Margrit Shildrick. 2002. "Bodies Together: Touch, Ethics and Disability." In *Disability/Postmodernity*, ed. Mairian Corker and Tom Shakespeare, 62–75. London and New York: Continuum.

Probyn, Elspeth. 1993. *Sexing the Self: Gendered Positions in Cultural Studies*. London and New York: Routledge.

———. 1996. *Outside Belongings*. New York and London: Routledge.

———. 2005. *Blush: Faces of Shame*. Minneapolis: University of Minnesota Press.

Radley, Alan. 2002. "Portrayals of Suffering: On Looking Away, Looking At, and the Comprehension of Illness Experience." *Body and Society* 8, no. 1: 1–23.

Radner, Gilda. 1989. *It's Always Something*. New York: Simon and Schuster.

Rajchman, John. 1999. "Diagram and Diagnosis." In *Becomings: Explorations in Time, Memory, and Futures*, ed. Elizabeth Grosz, 42–54. Ithaca, N.Y.: Cornell University Press.

Raup, R.M. 1959. *The Indian Health Program from 1800 to 1955*. Washington, D.C.: Department of Health, Education and Welfare.

Reason, James. 1990. *Human Error*. Cambridge: Cambridge University Press.

Reeve, Christopher. 1998. *Still Me*. New York: Random House.

Reiser, Stanley Joel. 1978. *Medicine and the Reign of Technology*. Cambridge: Cambridge University Press.

Rich, Frank. 1995. "Dance of Death." *New York Times*, 8 January, Section 1, 19.

Rieff, David. 2005. "Illness as More than Metaphor." *New York Times Magazine*, December 4.

Roeper, Richard. 1999. "Beware of Society's New Breed of Hatemongers." *Chicago Sun-Times*, 17 May, 11.

Rollin, Betty. 1976. *First, You Cry*. Philadelphia: Lippincott.

Rom, Mark C. 1997. *Fatal Extraction: The Story Behind the Florida Dentist Accused of Infecting His Patients with HIV and Poisoning Public Health*. San Francisco: Jossey-Bass Publishers.

Rose, Gillian. 1995. *Love's Work: A Reckoning with Life*. New York: Schocken Books.
———. 1996. *Mourning Becomes the Law: Philosophy and Representation*. Cambridge: Cambridge University Press.
Rose, Jacqueline. 1993. *Why War?* Oxford: Blackwell Publishers.
Rose, Nikolas. 1994. "Medicine, History and the Present." In *Reassessing Foucault: Power, Medicine, and the Body*, ed. Colin Jones and Roy Porter, 48–72. London and New York: Routledge.
———. 1996. "Identity, Genealogy, History." In *Questions of Cultural Identity*, ed. Stuart Hall and Paul Du Gay, 128–50. London: Sage Publications.
Rothman, Sheila M. 1994. *Living in the Shadow of Death: Tuberculosis and the Social Experience of Illness in American History*. New York: Basic Books.
Ruzek, Sheryl Burt. 1978. *The Women's Health Movement: Feminist Alternatives to Medical Control*. New York: Praeger.
Sacks, Oliver. 1983. *Awakenings*. New York: Dutton.
———. 1993 [1984]. *A Leg to Stand On*. New York: HarperCollins.
St. Pierre, Mark. 1991. *Madonna Swan: A Lakota Woman's Story*. Norman: University of Oklahoma Press.
Saywell, Cherise, Liza Beattie, and Lesley Henderson. 2000. "Sexualized Illness: The Newsworthy Body in Media Representations of Breast Cancer." In *Ideologies of Breast Cancer*, ed. Laura Potts, 37–62. New York: St. Martin's Press.
Scannell, Kate. 1999. *Death of the Good Doctor: Lessons from the Heart of the AIDS Epidemic*. San Francisco: Cleis Press.
Scarry, Elaine. 1985. *The Body in Pain: The Making and Unmaking of the World*. New York: Oxford University Press.
Schutz, Alfred, and Thomas Luckmann. 1973. *The Structures of the Life-World*. Trans. Richard Zaner and Tristram Engelhardt. Evanston, Ill.: Northwestern University Press.
Scott, Joan W. 1992. "'Experience.'" In *Feminists Theorize the Political*, ed. Judith Butler and Joan W. Scott, 22–40. New York and London: Routledge.
Seaton, Matt. 1999. "The Quick-Change Artist." *New York Times Book Review*, 29 August, 8.
Sedgwick, Eve Kosofsky. 1993a. "Queer Performativity: Henry James's *The Art of the Novel*." *GLQ* 1, no. 1: 1–16.
———. 1993b. *Tendencies*. Durham, N.C.: Duke University Press.
———. 1999. *A Dialogue on Love*. Boston: Beacon Press.
Sedgwick, Eve Kosofsky, and Adam Frank. 1995. "Shame in the Cybernetic Fold: Reading Silvan Tomkins." In *Shame and Its Sisters: A Silvan Tomkins Reader*, ed. Eve Kosofsky Sedgwick and Adam Frank, 1–31. Durham, N.C.: Duke University Press.
Segrave, Elisa. 1995. *The Diary of a Breast*. London and Boston: Faber and Faber.
Shakespeare, William. 1972. *King Lear*. Ed. Kenneth Muir. London: Methuen.
Shenk, David. 2001. *The Forgetting: Alzheimer's: Portrait of an Epidemic*. New York: Doubleday.
Shildrick, Margrit. 1997. *Leaky Bodies and Boundaries: Feminism, Postmodernism and (Bio)Ethics*. London and New York: Routledge.
Shildrick, Margrit, and Janet Price, eds. 1998. *Vital Signs: Feminist Reconfigurations of the Bio/logical Body*. Edinburgh: Edinburgh University Press.

Silversides, Ann. 2003. *AIDS Activist: Michael Lynch and the Politics of Community*. Toronto: Between the Lines.

Singer, Linda. 1993. *Erotic Welfare: Sexual Theory and Politics in the Age of Epidemic*. Ed. Judith Butler. New York and London: Routledge.

Sontag, Susan. 1966. *Against Interpretation and Other Essays*. New York: Anchor Books.

———. 1978. *Illness as Metaphor*. New York: Farrar, Straus and Giroux.

———. 1988. *AIDS and Its Metaphors*. London: Penguin Books.

Sophocles. 1957. *Antigone*. In *Ten Greek Plays in Contemporary Translations*. Trans. Shaemas O'Sheel. Ed. L. R. Lind, 77–109. Boston: Houghton Mifflin Company.

Spears, Ellen Griffith. 1998. *The Newtown Story: One Community's Fight for Environmental Justice*. Atlanta: The Center for Democratic Renewal and the Newtown Florist Club.

Spelman, Elizabeth. 1997. *Fruits of Sorrow: Framing Our Attention to Suffering*. Boston: Beacon Press.

Spivak, Gayatri Chakravorty. 1994. "Can the Subaltern Speak?" In *Colonial Discourse and Post-Colonial Theory: A Reader*, ed. Patrick Williams and Laura Chrisman, 66–111. New York: Columbia University Press.

Stabiner, Karen. 1997. *To Dance with the Devil: The New War on Breast Cancer*. New York: Dell.

Stacey, Jackie. 1997. *Teratologies: A Cultural Study of Cancer*. London and New York: Routledge.

Starr, Paul. 1982. *The Social Transformation of American Medicine: The Rise of a Sovereign Profession and the Making of a Vast Industry*. New York: Basic Books.

Steingraber, Sandra. 1997. *Living Downstream: A Scientist's Personal Investigation of Cancer and the Environment*. New York: Vintage.

Sullivan, Andrew. 1999. "There Will Always Be an England." *New York Times Magazine*, 21 February, 38–45; 54; 70–73; 78–79.

Sutherland, John. 2001. "This Man Is Irresistible to Younger Women." *Guardian*, 14 May, 4.

Swarns, Rachel L. 1999. "Writer Helps Soweto Strip the Shame from AIDS." *New York Times*, 24 October, 1 and 12.

Taylor, D. J. 2000. "Isn't He Rather Young to Be Writing His Memoirs?" *Independent*, 12 May, 5.

Thomson, Rosemarie Garland. 1996. *Freakery: Cultural Spectacles of the Extraordinary Body*. New York: New York University Press.

———. 1997. *Extraordinary Bodies: Figuring Physical Disability in American Culture and Literature*. New York: Columbia University Press.

Tilberis, Liz. 1998. *No Time to Die*. Boston: Little, Brown.

Toombs, S. Kay. 1992. "The Body in Multiple Sclerosis: A Patient's Perspective." In *The Body in Medical Thought and Practice*, ed. Drew Leder, 127–37. Dordrecht: Kluwer Academic Publishers.

Treichler, Paula. 1993. "AIDS Narratives on Television: Whose Story?" In *Writing AIDS: Gay Literature, Language, and Analysis*, ed. Timothy F. Murphy and Suzanne Poirier, 161–99. New York: Columbia University Press.

———. 1999. *How to Have Theory in an Epidemic: Cultural Chronicles of AIDS*. Durham, N.C.: Duke University Press.

Tronto, Joan. 1993. *Moral Boundaries: A Political Argument for an Ethic of Care*. New York: Routledge.

Verghese, Abraham. 1994. *My Own Country: A Doctor's Story of a Town and Its People in the Age of AIDS*. New York: Simon and Schuster.

———. 1998. *The Tennis Partner: A Doctor's Story of Friendship and Loss*. New York: HarperCollins.

Walker Art Center. 1998. *Art Performs Life: Merce Cunningham/Meredith Monk/Bill T. Jones*. Minneapolis, Minn.: Walker Art Center.

Ward, Diana. 1997. "The Women's Environmental Network: Mapping Environmental Causes of Breast Cancer." *Women's Health Journal* (July), 41.

Watt, Ben. 1996. *Patient: The True Story of a Rare Illness*. London: Viking.

Weeks, Jeffrey. 1989. "AIDS, Altruism and the New Right." In *Taking Liberties*, ed. Erica Carter and Simon Watney, 127–32. London: Serpent's Tail.

Wendell, Susan. 1996. *The Rejected Body: Feminist Philosophical Reflections on Disability*. New York and London: Routledge.

White, Ryan, and Ann Marie Cunningham. 1991. *Ryan White, My Own Story*. New York: Dial Books.

Whitehead, Tony. 1989. "The Voluntary Sector: Five Years On." In *Taking Liberties*, ed. Erica Carter and Simon Watney, 107–11. London: Serpent's Tail.

Wilber, Ken. 1991. *Grace and Grit: Spirituality and Healing in the Life and Death of Treya Killam Wilber*. Boston: Shambhala.

Williams, Terry Tempest. 1992. *Refuge: An Unnatural History of Family and Place*. New York: Vintage.

Williamson, Judith. 1989. "Every Virus Tells a Story: The Meaning of HIV and AIDS." In *Taking Liberties*, ed. Erica Carter and Simon Watney, 69–80. London: Serpent's Tail.

Wilton, Tamsin. 1992. *Antibody Politic: AIDS and Society*. Gloucester: New Clarion Press.

Wingo, Phyllis A., Eugenia E. Calle, and Anne McTiernan. 2000. "How Does Breast Cancer Mortality Compare with That of Other Cancers and Selected Cardiovascular Diseases at Different Ages in U.S. Women," *Journal of Women's Health and Gender-Based Medicine* 9, no. 9: 1004.

Winnicott, D. W. 1971. *Playing and Reality*. London and New York: Routledge.

Wittman, Juliet. 1993. *Breast Cancer Journal: A Century of Petals*. Golden, Colo.: Fulcrum Publishing.

Yalom, Marilyn. 1997. *A History of the Breast*. New York: Knopf.

Yingling, Thomas. 1991. "AIDS in America: Postmodern Governance, Identity, and Experience." In *Inside/Out: Lesbian Theories, Gay Theories*, ed. Diana Fuss, 291–310. New York and London: Routledge.

Young, Iris Marion. 1990. *Throwing Like a Girl and Other Essays in Feminist Philosophy and Social Theory*. Bloomington: Indiana University Press.

Zeiger, Melissa F. 1997. *Beyond Consolation: Death, Sexuality, and the Changing Shapes of Elegy*. Ithaca, N.Y.: Cornell University Press.

Zizek, Slavoj. 1989. *The Sublime Object of Ideology*. London and New York: Verso.

———. 1991. *Looking Awry: An Introduction to Jacques Lacan through Popular Culture*. Cambridge, Mass.: MIT Press.

Zola, Irving Kenneth. 1983. *Socio-Medical Inquiries: Recollections, Reflections, and Reconsiderations*. Philadelphia: Temple University Press.

Zuger, Abigail. 1995. *Strong Shadows: Scenes from an Inner City AIDS Clinic*. New York: W. H. Freeman and Company.

Index

Lisa Diedrich is assistant professor of women's studies at Stony Brook University.